2D

A-Z
BIOCHEMISTRY

A-Z BIOCHEMISTRY

Dr. Albert Shawn

CENTRUM PRESS
NEW DELHI-110002 (INDIA)

CENTRUM PRESS
H.O.: 4360/4, Ansari Road, Daryaganj,
New Delhi-110 002 (India)
Ph.: 23278000, 23261597
B.O.: No. 1015, Ist Main Road, BSK IIIrd Stage
IIIrd Phase, IIIrd Block,
Bangalore - 560 085 (India)
Tel.: 080-41723429
Visit us at: www.centrumpress.com

A-Z Biochemistry

First Edition, 2009

ISBN 978-93-80106-05-2

PRINTED IN INDIA

Printed at Salasar Imaging Systems, Delhi-110035 (India)

Contents

Preface

Biochemistry is a *Hybrid Science*: Biology is the science of *Living Organisms and Chemistry* is the *Science of Atoms and Molecules,* so biochemistry is the science of the atoms and molecules in living organisms. Its domain encompasses the entire living world with the unifying interest in the *Chemical Structures* and reactions that occur in living systems. It can also be defined as the chemistry of Biology. The science of biochemistry has also been called *Physiological Chemistry and Biological Chemistry.*

Biochemistry is concerned chiefly with the chemistry of *Biological Processes*; it attempts to utilize the tools and concepts of chemistry particularly organic and physical chemistry, for elucidation of the living system. The Biochemist studies *biological systems* at the level of Molecules and Atoms, the level of Molecular Biology. Biochemical approaches to the simplification and understanding of biological systems require two types of background.

First, biochemists must be thoroughly skilled in the *Basic Principles and Techniques of Chemistry, such as Stoichiometry, Photometry, Organic Chemistry, Oxidation and Reduction, Chromatography, and Kinetics.* Second, Biochemists must be familiar with the theories and principles of a wide variety of Biological and Physical disciplines often used in Biochemical studies, such as *Genetics, Radioisotope Tracing, Bacteriology, and Electronics.* This need reflects the Biochemists' ready acceptance and use of theories and techniques from allied areas and disciplines.

Author

Chapter 1

Constituents of Living Things

The world, in itself, is a colourless, soundless, smell-less affair which none the less manages to titillate the senses so that they take on the garb of the more flashy secondary qualities. The intellectual tension caused by splitting and pulling apart the primary from the secondary qualities can be seen in the work of that honest underlabourer for science, John Locke. We can see the problem arising if we ask the question: What distinguishes atoms from the empty space they are supposed to occupy? Locke, closely following Robert Boyle, maintains that atoms possess primary qualities only. What are these qualities? Locke's list varies somewhat but we can take the following as typical: 'These I call *original or primary qualities* of body, which I think we may observe to produce simple ideas in us, viz. solidity, extension, figure, motion or rest, and number'.

Now if we look closely at this list then it seems evident that solidity is the only one of the primary qualities capable of doing the job of differentiating atoms from space. But what *is* solidity? He distinguishes it from hardness by saying 'that solidity consist in repletion, and so an utter exclusion of other bodies out of the space it possesses'; but, of course, to be replete is to be stuffed with, or full of, something or other and we still do not know what this something or other *is*.

The trouble now is (if we stay within Locke's scheme of primary and secondary qualities and within his empiricism) that the only kind of qualities with which we are directly acquainted and which are space-filling just are the secondary qualities. Colours and warmths, for example, occupy definite positions and take up shapes in space. In fact, Locke tries to save his empiricism by illicitly making the notion of solidity straddle the divide, on the one hand, between primary and secondary qualities and, on the other, between a directly perceivable quality and one that is not perceivable but which *explains* the space-filling and causal properties of bodies.

We can see this tension right at the beginning of his extended discussion of the 'simple idea' (that is, the simple sensation) of solidity: 'The idea of *solidity* we receive by our touch; and it arises from the resistance which we find in body to the entrance of any other body into the place it possesses till it has left it'.

Already we can see him mixing up talk of the sensation of touch (what we might call a solidity-sensation) with what *causes* that sensation (resistance). (And, of course, if we are realists we should not accept that we get the notion of one body resisting another only from touch: we can, for example, just as well *see* bodies resisting each other). And at the end of the same section he makes the move of claiming that the purported explanatory notion of solidity is the very same as, or is very similar to, the simple *idea* of solidity which we get through touch:

And though our senses take no notice of it but in masses of matter, of a bulk sufficient to cause a sensation in us, yet the mind, having once got this idea from such grosser sensible bodies, traces it further, and considers it, as well as figure, in the minutest particle of matter that can exist, and finds it inseparably inherent in body, wherever or however modified.

But the simple idea of solidity cannot be the same as the explanatory notion. The *idea* is just that, one of several, independent ideas or sensations whereas the *notion* of solidity is meant to explain several things about matter. And so he tells us:

By this idea of solidity is the extension of body distinguished from the extension of space, the extension of body being nothing but the cohesion or continuity of solid, separable, movable parts; and the extension of space the continuity of unsolid, inseparable, and immovable parts. Upon the solidity of bodies also depend their mutual impulse, resistance and protrusions.

This 'quality' of solidity obviously is not gained directly from experience and the pressing questions arise of what it possibly can be and how we can get to know what it is.

DISCOURSE AND REALITY LEVELS

Atomic theory, or its near relation, molecular biology, now claims to be able to give, in principle, a complete reductive account of sentient organisms. We are, in the first analysis, not much more than complex bundles of nucleic acid and protein and, in the last analysis, mere structured heaps of quarks and leptons. Many people have, naturally enough, disliked the idea that there is no place in the world for colours, scents and so forth and have liked even less the thought that nowhere, not even in their own heads, is there a mind to be found.

A move, popular among a number of biologists and philosophers who have reflected on these issues, has been to introduce the notion of *levels*. In this way they hope to retain realism about science while, at the same time, escaping reductionism and its implication. Thus, it might be said, although it is true that at one level, organisms are composed of quarks and leptons, it is also true that, at a different level, they are made of cells and, at a different level still, they may be conscious beings.

Very often such claims are combined with a notion of 'emergent evolution' in which it is held that quite new properties, such as consciousness, come into existence with, say, the increasing complexity of things. In any case, it is important to notice that these different levels—in which what gets squeezed out at a lower level may be restored at a higher—are taken as having independent, real existences. There is no question of the reduction, even in principle, of upper levels to lower.

It should at once be remarked that the ontological reductionist is not refuted by the fact that biologists, in their explanations, often mix descriptions and theories from different levels. It does not follow from the fact that the mixing of levels of *discourse* is frequent, and often necessary, that there must, therefore, be different *ontological* levels. James J. Gibson has tried to refute those reductionists who, 'impressed by the success of atomic physics, have concluded that the terrestrial world of surfaces, objects, places, and events is a fiction'. His answer to this is:

The world can be analyzed at many levels, from atomic through terrestrial to cosmic. There is physical structure on the scale of millimicrons at one extreme and on the scale of light years at another. But surely the appropriate scale for animals is the intermediate one of millimetres to kilometres, and it is appropriate because the world and the animal are then comparable.

The reductionist's reply to this will surely be that mention of the supposed features of macroscopic objects, in explaining animal behaviour, is simply a methodological convenience— or even a methodological necessity.

For example, in the explanation of visual perception statements about the patterns of light and shade on objects may be mixed in with statements, at a different level, concerning the cells of the retina and the brain and yet others, at still another level, about photons. But one may take advantage of this methodological mix without being committed to the real existence of all the supposed levels.

Similarly, there is more to the reductionist's case than justifies its dismissal, as Max Deutscher has recently undertaken, as a piece of obsessional thinking. Deutscher claims that the answer to the 'feeling' of reductive materialism 'lies in reminding ourselves of the plurality of facts and conceptual forms, and in regaining a free capacity to shift our perspectives'.

As barely stated this objection merely begs the question because, whatever may be the case concerning the plurality of conceptual forms or perspectives, the reductive materialist's argument just is that, in the sense required, there is *no* plurality of facts. As Stephen W. Hawking has put it: 'Since the structure of molecules and their reactions underlie all of chemistry and biology, quantum mechanics allows us in principle to predict nearly everything we see around us, within the limits set by the uncertainty principle.'

It would, says Deutscher, be as difficult 'to bring a volcano as to bring consciousness into a picture which allows no character or description except that contained within basic physics'. No doubt; but this incapacity, by itself, does not show there is anything more to a volcano than a collection of fundamental particles. It does not force the conclusion that:

It is foolish to deny that volcanoes are composed of the elementary particles, and equally foolish to suppose that a story in terms of elementary particles presents the reality of things, whereas a story in the large scale about volcanoes and lava flows presents only a superficial appearance of the world.

Part of the problem here is that Deutscher does not spell out his 'story in the large scale'; and a reductive materialist would surely claim that the *full* story in the large scale, if only it were possible to tell it, would be in terms of the fundamental particles which together make up the volcano. As a matter of fact, scientists do try to give as fully a reductive account as possible of matter 'in the large' and in some cases, as with the quantum mechanical explanation and prediction of the properties of crystals and fluids, they have achieved remarkable success.

This last point deserves further emphasis because it also has force against Popper's claim that there have been very few successes for the proposed reductionist programme. The reductionist can, of course, here simply fall back on the fact of the great complexity of many things in the world. This is a hindrance to a full reductionist account being given of organisms, for example.

But the reductionist can go further than this and point to the many successful achievements of the quantum mechanical theorists in recent years. These theorists have been able to explain and predict the properties of many materials, for example, because they have found ways of making simplifying assumptions concerning their constituent atoms.

These simplifying assumptions, themselves, depend on the presupposition that a full account of materials does depend on quantum-mechanical principles. Thus the wave functions of the all-important valence-electrons of an atom can be calculated to a good approximation by treating the nucleus and the inner electron shells as together constituting a 'pseudoatom' and this is an important step on the way to predicting important properties of crystals such as gallenium sulphide. As Cohen *et al.*, from whom I take this account, put it:

Today the quantum mechanics of materials has become both conceptually and practically a simpler study than the study of the electronic structure of atoms having more than one possible valence-electron configuration. Through worldwide collaboration on pseudoatoms, alchemy, the original black art of materials science, is now well on its way to becoming one of the better developed parts of human knowledge.

Similar remarks could also be made about the theoretical and practical investigations of fluids. Unfortunately, the point is obscured by some authors who present us with a mishmash of levels of description and levels of actuality. Thus Steven Rose, partly to allay our fears about the implications of reductionism for the dignity of human beings, tells us 'that there are many levels at which one can describe the behaviour of the brain'; but he vacillates between talking about levels of description, which are in different 'universes of discourse', and discussing whether or not there can be causal relations 'between the point-set on one level and that on another'.

The way he answers the latter question, by suggesting that we should be cautious in attributing causal rather than mere correlative relationships, suggests that he is now speaking of ontological levels so that, for example, the state of being in love occurs on one level and is correlated with molecular changes on another.

On the other hand, it seems that for Kathleen Wilkes there could be no question of a confusion of levels of language and ontological levels because ontological questions just *are* questions about language. In the midst of her spirited defence of physicalism about the mind-body problem she suddenly admits 'that the ontology, terms and explanations of common sense have nothing to do with physicalism'.

ONTOLOGICAL LEVELS

When we come to examine the notions of those thinkers who clearly espouse the doctrine of different levels there are further problems. There are variations in what are supposed to exist at different levels—qualities, laws, 'principles', things all have their advocates—and it may even be the case that, in a given hierarchical schema, different kinds of 'entities' occupy the various levels, e.g., things at one level and 'principles' at another. If we leave this problem aside there are still often difficulties in the interpretation of the concrete examples that various thinkers give. Thus consider the hierarchical list of levels proposed by Popper:

- Level of ecosystems
- Level of populations of metazoa and plants
- Level of metazoa and multicellular plants
- Level of tissues and organs (and of sponges?)
- Level of populations of unicellular organisms
- Level of cells and of unicellular organisms
- Level of organelle (and perhaps of viruses)
- Liquids and solids (crystals)
- Molecules
- Atoms
- Elementary particles
- Sub-elementary particles
- Unknown-sub-sub-elementary particles?

It is as if, with wedges and stars, the 'whole structure' of these things floats free from the fundamental constituents and, from its superior height, acts down upon them. I, for one, cannot make sense of this supposition.

We can see a related confusion in the work of those thinkers who wish to defend the notion of an infinite number of levels of physical being. David Bohm, for example, has urged that the history of atomic physics suggests that just as beneath the level of macroscopic objects there lies the level of atoms and molecules and beyond that again the level of 'fundamental particles', so, at least in principle, the physicist might go on and on discovering, without end, level after level.

Now I think that in some sense the universe may be infinitely complex, and that, because of this, scientists may find eternal employment in unravelling the complexity; and it may also be that, despite their efforts, scientists will never get down to ontological bedrock; but it does not follow from this, nor indeed does the view make much sense, that there could exist an infinite number of ontological levels or strata. Indeed, I do not think there could exist more than one such level.

Once more the spatial metaphor of 'levels' in discussions of ontology is highly misleading. It is not the case, for example, that there is one set of entities, the protons, and *also* existing 'beneath' it, in some strange ontological space, a swarming sea of quarks which somehow supports its protonic flotsam and jetsam. The scientist, in finding out that protons are *really* triads of quarks in close interaction is discovering that there are no such separate things as protons—or, what comes to the same thing, that protons, in being made up of triads of mutually bound quarks, just *are* those threesomes in bondage.

To believe otherwise would, if the belief were rigorously followed through, involve awkward questions about the causal interactions between the protons and their 'constituent' quarks, remembering that the existence of the latter was, in the first place, postulated in order *fully* to explain the behaviour of those clumpings in space-time that were originally called 'protons'. Even if it were found that the quark theory did not provide a full explanation, that, for example, it were necessary to postulate some kind of 'membranes' to contain the separate triads of quarks, it would still be the case that the quarks and the 'membranes' all exist in causal interaction at the one level.

The attempt, then, to postulate an infinite series of levels of physical existence would be to believe in an infinite series of 'seemings to be' perched upon nothing at all.

POLANYI'S NOTION OF LEVELS

The reductionist's ability plausibly to deflect apparent counter-examples into the basket of merely methodological issues must strengthen his ontological claims. In fact, it is possible to show that in one recent influential positive defence of the idea of levels, that of Michael Polanyi, the argument depends for its apparent strength on an illicit mixing of ontological, and what may be loosely called methodological, claims. Because of this and other confusions the argument is difficult to state concisely. For convenience, I will first use Marjorie Grene's very clear formulation.

Since they declare the processes of growth and heredity have been shown to be determined by a sequence of DNA molecules, biology has already been reduced, on principle, to bio-chemistry; the completion of the job is routine. Granted, however, that DNA has precisely the power they claim for it, its operation demonstrates, on the contrary, that biology is *not* reducible to biochemistry (and ultimately to physics).

What makes DNA do its work is not its chemistry but the order of bases along the DNA chain. It is this order which functions as a code to be read out by the developing organism. The laws of physics and chemistry hold, as reductivists rightly insist, universally; they are entirely unaffected by the particular linear sequence that characterizes the triplet code. Any order is possible physico-chemically; therefore physics and chemistry cannot specify *which* order will in fact succeed in functioning as a code.

This argument, which Grene thinks 'appears incontrovertible', would, if valid, beard the reductionist lion in his den. In fact, the argument very clearly is fallacious. If we look closely at the last sentence quoted above we perceive an equivocation. When Grene says, of the bits that make up the DNA molecule, that any order is possible physico-chemically she is thinking, as in the preceding sentence she states, of the *laws* of physics and chemistry; and scientific laws, by their nature merely allow *possibilities*—they do not, by themselves and in the absence of a statement of initial conditions, state what *must* happen.

When she concludes that 'physics and chemistry cannot specify *which* order will in fact succeed in functioning as a code' then, if her argument is to bear on the reductionist claim, she must now be thinking of the actual entities, including forces etc., that chemistry and physics describe.

The conclusion must be that the structure of a particular DNA molecule is not due to the way its constituents as originally independent entities bumped up against, and reacted with, each other and this is borne out by the way Polanyi puts it: 'As the arrangement of a printed page is extraneous to the chemistry of the printed page, so is the base sequence in a DNA molecule extraneous to the chemical forces at work in the DNA molecule'. Because of this equivocation, the conclusion does not follow.

Once we do not feel forced to swallow the conclusion, because of the apparent reasonableness of the argument that led to it, we may feel free to examine it in its own right. When we do this we surely must see just how outrageous Polanyi's view is. The correct view would seem to be that it is just because the constituents of DNA are what they are, and because of the forces between them, that the laws of physics and chemistry 'are entirely unaffected by the particular linear sequence that characterizes the triplet code', i.e. that these laws 'allow' any *linear* sequence while ruling out other configurations.

The virtue of Grene's formulation of Polanyi's argument is that it allows us very quickly to see what is wrong with it. However, as Polanyi puts it, the argument comes embedded in a longer discussion, where there is a confusing mixture of ontological and 'methodological' issues, and itself then takes over this same confusion. It is precisely this that makes the argument seem so plausible.

There is a hint of what is happening in Grene's account when she says, 'What makes DNA do its work is not its chemistry but the order of bases along the DNA chain. It is this order which functions as a code to be read out by the developing organism.' One's immediate response to this is that there is no distinction between the chemistry of the DNA molecule and the order of the bases any more than one distinguishes between the chemistry of acetic acid and the disposition of its component atoms.

One might say the structure of the molecule *is* part of its chemistry. The fatally confused step is when Grene goes on to say that the order of bases 'functions as a code' as if acting as a code involved something over and above acting chemically. In fact, her talk of the code being 'read out by the developing organism' should not be taken as a mere rhetorical flourish, as one might imagine it being used by a hard-line reductionist like Francis Crick, but is precisely where the whole confusion lies. This comes out in Polanyi's exposition where he talks at first of the DNA code which has 'information content' that, as he makes clear, is given by 'the numerical improbability of the arrangement'; but then immediately goes on to change his mind as to what 'information content' really is.

A printed page may be a mere jumble of words, and it has then no information content. So the improbability count gives the *possible,* rather than the *actual,* information content attributed to a DNA molecule; the sequence of the bases is deemed meaningful only because we assume with Watson and Crick that the arrangement generates the structure of the offspring by endowing it with its own information content.

Polanyi has gone from speaking of 'information content' in a technical sense, as used in information theory, to a quite different use more akin to 'meaning' as when we speak of the meaning of a piece of prose. It is, of course, the first sense which is used in biological *science* and will be insisted upon by the reductionist as posing no threat to his reductionism.

This blatant mystery-mongering has its purpose. For Polanyi wishes to regard the DNA molecule as a mixture of blueprint and engineer which somehow *constructs* the living organism. 'Can the control of morphogenesis by DNA be likened to the designing and shaping of a machine by the engineer?' he asks, and answers in the affirmative.

The point of this comparison is that he has earlier likened organisms to machines and has tried to show that even with ordinary machines, such as clocks, a reductive account of them cannot be given.

This is because a machine 'works under the control of two distinct principles. The higher one is the principle of the machine's design, and this harnesses the lower one, which consists in the physical-chemical processes on which the machine relies'.

We might be tempted to regard the higher principle, that of the machine's design, which he describes as a 'boundary condition', as merely referring to the design of the machine as it is embodied in the machine, i.e. as the machine's structure; and then talk of the lower principles, i.e. the physical-chemical processes, as being 'harnessed' by the higher is at most a misleading metaphor.

In one mood he does speak in this way as when he says, This harness is not unbreakable; the structure of the machine, and thus its working, can break down. But this will not affect the forces of inanimate nature on which the operation of the machine relied; it merely releases them from the restriction the machine imposed on them before it broke down.

If we are thinking, in this way, of the machine, in itself, then its structure does not harness its matter as a rider harnesses a horse—this sort of language can only be a florid way of saying that the structural arrangement of the material has to be given as an initial condition before, from it and the laws of mechanics etc., we can say what shall happen. There is no question here of higher and lower 'principles' and, therefore, no threat to the reductionist's position.

This is disguised by the fact that Polanyi does not consistently refer to the machine and its structure, in itself, but also brings in the idea of the *constructor* of the machine. It is not now merely a question of the machine's structure harnessing the laws of nature—an innocuous idea if interpreted correctly—but the quite different concept of an *engineer* harnessing them.

The structure of machines and the working of their structure are thus shaped by man, even while their material and the forces that operate them obey the laws of inanimate nature. In constructing a machine and supplying it with power, we harness the laws of nature at work in its material and in its driving force and make them serve our purpose. Of course, in a sense, the engineer harnesses the laws of nature by bringing bits of matter into certain relations. But in saying this we are no longer thinking of higher and lower principles 'in' or 'of' the machine which is the conclusion Polanyi desires. If we are to speak, still, of two principles then it should be clear that there is now only one principle for the matter which makes up the machine and that the other resides with the engineer, which is only a complicated way of saying that there are two things in a certain relation to each other. Once seen in this light there seems no reason why a reductive account should not be given of the complex system: engineer plus machine.

In a loose way we can say that there is a confusion of ontological with methodological questions in that bringing in the engineer brings in a consideration of *interests*. The engineer, we might say, has a particular *interest* in the structure of a machine, either in bringing it about or in merely studying it. Similarly, a chemist may be interested in the atoms of which it is made.

In fact, Polanyi, when speaking of a distinction he makes between machine type boundaries and test-tube type boundaries, seems to distinguish between the two according to whether our interest is in the boundary or the matter that is bounded; and that it is a purely methodological distinction is further suggested by his remark that 'By shifting our attention, we may sometimes change a boundary from one type to another'.

Also, in his *Personal Knowledge,* in what may plausibly be seen as a fore-runner of the argument here criticised, Polanyi distinguishes between the physics and chemistry of machines and their 'operational principles', the latter being '*rules of rightness,* which account only for the successful working of machines but leave their failures entirely unexplained'. Here, again, I think we can easily detect questions of interest being confused with questions concerning the fundamental nature of things but I shall not pursue the issue further.

THE CONTRADICTION IN THE NOTION OF LEVELS

To assert the existence of different levels of 'principles' etc., unless this is taken as an obscure way of referring to different things interacting in a common environment, is to invite a criticism of the kind used to effect by the late Professor John Anderson and dubbed the 'two worlds argument' by John Passmore. This is that once two ontological levels, two distinct kinds of being, are distinguished then there is no way that they can, without contradiction, be brought together again in mutual interactions. The power of Anderson's argument is evident as it is employed against such metaphysical ideas as the ontological distinction between the platonic forms and the world of becoming or between God as the wholly other and the finite world he creates.

However, it can also be used where the proposed ontological 'split' is not, at least at first sight, as wide as this—including Polanyi's hierarchies if we take these to be ontological. For now it seems that in machines, for example, with the operation of higher order principles, the matter of which the machines are composed must act out of character.

This is so because, in not taking these higher order principles as merely what occurs when bits of matter come into a certain relation with other things in the universe, e.g. engineers, they must be regarded as of the nature of those bits of matter.

By then saying that the lower order principles are 'open' to the higher, i.e. that they go on irrespective of the existence of the higher even although the latter also affect the action of the former, he is asserting that the matter of machines has both the character X and not-X.

The argument has been put by P.H. Partridge for the general notion of emergent evolution. After arguing that the concept of causality as creative emergent evolution must take its sense from a supposed contrast with a preformationist idea of causality, where the effect somehow lies within the cause, he continues:

The theory of preformationism simply means that the effect is deducible from the cause alone, or is necessitated by the cause, it being the nature of a particular cause to produce such an effect. If, then, it is the nature of A to produce the one effect B, when A 'transcends itself by producing C, it is not acting according to its nature, but according to some other nature that is not its. Hence, the assumption that a thing both necessitates and creates implies that it is both itself and not itself, and this difficulty can be met not by a revision of logic, but, if we are unwilling to give up the notion of emergence as a special and peculiar occurrence, by the invention of some higher, mythological power asserting itself through the cause.

If it is denied that the concept of emergent evolution involves the underlying assumption of these two different kinds of causality then the idea must be of one kind, which is creative in the sense that the effect is different from the cause, but with which one can nevertheless distinguish causes which have the most remarkable effects: effects which, as Samuel Alexander suggests, must be accepted with 'natural piety'.

There is here clearly a problem of how we distinguish between emergent and non-emergent effects although Alexander's own belief in the emergence of material things from naked space-time—as though space-time were a kind of unsuccessful porridge—may suggest there are some paradigms of the former. The point here seems to be that if such astonishing effects do burst forth in the cosmos then the rational course is not to accept them with 'natural piety' but to try to understand them and this involves finding out what their real causes are.

An illustration of this point is afforded by a defence of the idea of emergence by the British Communist, John Lewis. Consider his account of the evolution of the DNA molecule:

What has to be noted is that:

- The process is wholly explicable in a succession of physical states operating under physical laws which
- Bring about the complete novelty of the DNA molecule possessing new properties and operating under its own laws.

Certainly the DNA molecule is unique but only in the sense that it is a unique spatial arrangement of its constituent atoms which, given what they are, will, in the relevant circumstances, fall out in the order they do. In this sense, the water molecule is also unique and, indeed, any molecule we care to think of. Of course, *in a sense,* the DNA molecule possesses new properties but only in the sense that any molecule possesses new properties, that is, those properties fitting for the kind of thing it is; but this sense does not exclude the fact that these new properties are fully explained in terms of the constituent atoms including the spatial etc. relations between them. In fact, the chemistry of one virus, X174, has been worked out completely, and the sequence of 5,375 nucleotides along its DNA, and what it 'codes' for, is fully known. The DNA of X174 did spring some surprises, connected with the details of the coding, but these were ordinary chemical surprises. It did not, for example, suddenly start whistling 'Waltzing Matilda'.

Suppose we do accept the idea of new qualities in evolution, does not this, despite Lewis's disclaimers, make their emergence entirely mysterious? Here the dualist might find his mark.

Lewis claims that his scheme is not at all mysterious but this can only be in the sense that it is naturalistic and makes no appeal to 'supernatural' causes; but it certainly is mysterious in the sense that no *explanation* is possible of the emergent qualities nor, even, of why, for example, qualitatively identical DNA molecules should have the same emergent qualities.

Creative the DNA molecules may be but they have all the spontaneity of the Guards on parade. The dualists might claim that their own account in terms of a creative, spiritual mind is superior because it can explain the emergence of new qualities. Or it may be that the dualist, like the mechanist, will reject the idea of emergence but, unlike the latter, accept the existence of undeniably mental events irreducible to mechanistic explanations.

It is the existence of such events which leads the dualist to dualism. We bake a cake and, as it cools, from nowhere a deep red icing settles on its surface. How do we explain this? By denying that it is a *magic* cake and instead saying that it is a *unique* cake where the flour and the currants have settled in just that one combination that allows for the *emergence* of red icing? Is this any more mysterious than the DNA molecule from nowhere obtaining brand new properties or the brain a coating of mentality?

And if we try to lessen the mystery of the icing by postulating the existence of a ghostly chef, some departed Party member, perhaps, still anxious to do his bit, why would we think this more reasonable than conjuring up a god to paint on new qualities wherever in the universe they are needed?

If any content at all is to be given to ideas of 'emergence' or of 'hierarchies' then these ideas must be about what it is that exists at the one ontological level so that, for example, the dispute between the believer in emergent evolution and the mechanist is about whether or not vital forces or secondary qualities exist as well as quarks.

This can be seen in Polanyi whose talk of mysterious emergent higher principles at one point becomes deflected into the advocacy of vitalism. Here the dispute is with orthodox Darwinians where Polanyi argues that rather than studying the origins of new populations one should try to account for the evolution of a single man by referring to that man's family tree which, he claims, 'includes everything that has contributed to the maturing of this human being'.

The point here is that, by denying the importance of commonly accepted environmental factors in human evolution, involving natural selection etc., Polanyi is, naturally enough, obliged to invent a different environment containing vital forces in order to account for this evolution; but these vital forces can then no longer be regarded as having a special ontological priority but must simply be bits of the universe interacting with other bits and, in particular, the complex chemical systems which are living organisms. This view, once teased out from the confusing talk of 'hierarchies' or 'emergence' and stated bluntly in its own right, can then be more coolly assessed.

TENSIONS

Modern biology, then, involves two fundamental metaphysical assumptions. The first is that, despite the useful retention of teleological forms of language, all causes of biological functional activity are efficient causes; and the second is that what, ultimately, exist to be studied by biology are structured collections of fundamental particles and their associated fields—*and nothing more*.

There is the further issue of whether, as is sometimes suggested, the category of causality should be given up altogether but, leaving that aside, these two general views are logically compatible. However, when we turn to the special Darwinian explanation of the origins of living things-which, of course, has played some part in the downfall of strictly teleological accounts—there is the problem of whether *it* is compatible with the strongly reductionist accounts of biophysics. The question is not often asked but needs to be taken seriously.

The theory of natural selection is, or at least, as originally conceived, was, concerned largely with explanation at the 'common-sense', macroscopic level and there is no guarantee that it will not suffer the fate of other theories of this kind that usefully miss the mark. As with the mechanics of gross bodies, the theory may be retained while the truth of the matter lies elsewhere, in among the strange interactions of what there ultimately is according to the quantum physicists.

Modern biology has had much to say about the fundamental constitutions of living things, including humans, and a good deal of it has been strongly reductionistic. If we leave aside the attempts of the logical empiricists to find logical relations of the appropriate sort between *propositions*—e.g. between the propositions of classical Mendelian genetics and those of modern biochemical genetics—then it seems that roughly two kinds of ontological reductionism have occupied thinkers in this area.

On the one hand there are what we may generically refer to as naked ape types of theory and, on the other, there is the view of biochemists, such as Jacques Monod, that organisms (including humans) are nothing more than collections of molecules in certain complex relationships. Here the difference between the two kinds of theory appears as a difference in the level of explanation, taking this as relative to ordinary, everyday descriptions of the world.

Naked ape types of theory, including behaviouristic, socio-biological, etc. theories, remain close to everyday descriptions—it is just that one commonly noticed piece of behaviour is explained away as really being some other commonly noticed piece of behaviour as when it may be said that love is but lust with a fancy name. With the molecular biologists the world they describe seems quite different from the one that, in our non-speculative moments, we seem to inhabit. Colours, sounds, familiar objects all vanish.

If, for the time being, we accept this rough distinction between the two kinds of reductionist explanation, then we may notice that theories of the naked ape type—which try to pick on a few things to explain everything (both the few and the many things coming, in the first instance, from the level of the macroscopic)—are open to the sort of objection Robert Boyle used against the 'chymists' of his time. In the *Sceptical Chemist* Boyle argued against the then current theories of the 'elements' or 'principles' of matter (whether these be taken as earth, air, fire and water or, as in a rival scheme, salt, sulphur and mercury) that he could see no reason why these should be picked out of a myriad others, e.g. oils, as *the* elements.

Similarly, why should certain bits of behaviour, e.g. sexual behaviour, be picked out as *the* important bits when it is clear that there are many kinds of behaviour? By their very nature such theories are liable to leave important things unexplained. Boyle argued that the excessive narrowness of the chemical theories of his time, and their consequent lack of explanatory power, was heavily disguised by obfuscation of various sorts so that, for example, the chemists' elements took on all kinds of occult qualities.

We might expect, in our time, to detect similar techniques at work in behaviouristic theories, sociobiology, etc. where the existence of what does plainly exist, e.g. consciousness, altruism and so on, is either denied or in some way redefined so as not to constitute a threat to the theory.

Often enough the redefinition of ordinary terms may serve

to disguise various indecencies committed in the name of science, as when behaviourists use the term 'punishment' in a technical sense to designate actions which, in ordinary language, would properly be described as assault.

In particular, we should be on the lookout for the supposed basic elements of human nature taking on magical qualities so that, for example, the 'hunting instinct', supposedly exclusively possessed by men, becomes transmuted into the *naturally* male-dominated *pursuit* of knowledge; and where, with as much right to the carefree use of metaphor, feminists might claim that the *birth* and *nurture* of theories is an expression of the 'mothering instinct'. Such thinking, as many authors have pointed out, very often has a plainly ideological function.

In so far as a sharp distinction can be made between the different levels of theory such criticisms of naked ape type theories are just. However, the comparison with the theory of the elements suggests that a sharp distinction cannot be made between different levels of theory in this way if these are taken in relation to everyday descriptions.

What, on one view, might be seen as the attribution of merely occult qualities to the elements, in order to save appearances, may, on a different view, be more kindly regarded as an attempt to look behind the complex, changing world for ultimate principles of explanation.

On this account, the chemists of Boyle's time are squarely in the western intellectual tradition of trying to find what it is that lies beyond the world of appearances and, in some sense, explains that world. It is merely that, unlike the atomists, say, they picked on the wrong kinds of things or qualities for their explanations. Similarly, hypotheses of hunting instincts etc. which, from one point of view, seem to be magically transmuted into activities usually seen as quite distinct from hunting etc. behaviour, may also be seen as attempts to penetrate behind appearances.

The latter interpretation can be illustrated by reference to pre-Socratic thought where the attempt was first made to find the basic stuff or stuffs of the world. Thales suggested it was water, perhaps, as Aristotle suggests in the *Metaphysics*, 'from observing that the nutriment of everything is moist, and that even heat is generated from moisture upon which it depends for its continued existence.

We might guess, from this suggestion of Aristotle, that the important transition in Thales's thought is from the more primitive question, 'Where do the things of the world come from?' which was hitherto answered by means of stories about the gods, to the question, 'What are the basic elements, or what is the basic stuff, of things?' If one asks the latter question then a satisfactory answer to it must involve choosing a candidate that, first, can be shown to be ubiquitous, to be in all things, and, second, has the right properties to explain the apparent diversity of things.

That water satisfied the first condition might have been suggested by what Aristotle gives as Thales's second reason for picking on it, 'that the seeds of everything have a moist character, and that water is the first principle in the character of all moist things'. Perhaps Thales thought the second condition was satisfied partly because he thought that one could observe, as a matter of fact, creatures emerging from watery substances such as the sea or semen; but partly, also, because he knew that water existed as various *phases* (liquid, solid and gas) and so, to that extent, could explain the apparent diversity of things in the world.

If this account of what Thales was about is anywhere near the truth then we can see that, like the seventeenth-century chemists or the twentieth-century naked ape theorists, he is picking on some one kind of thing to explain the characteristics of a multitude of things; and that, by doing this, he founds that intellectual adventure which endeavours to discover what lies behind the many perceived things.

To pick on water as the basic stuff of the world is, of course, to lay oneself open to the criticism, used by Thales's successors, that it simply will not explain the great diversity of objects. However, we can see with Thales two characteristic features of reductionist theories.

The first is that the entities referred to by the reducing

theory are picked as having certain general characteristics, e.g. the capacity of water to form the three phases of solid, liquid and gas, by which the entities referred to by the reduced theory or description are accounted for.

The second and correlative feature is that some characteristics only of the world that is said to be reduced are directly explained—thus Thales's theory may be said to explain, in a general way, why the world is made up of solids, liquids and gases but not the particular properties of, say, rocks, alcohol and mist. These two features stand in a potentially dialectical relationship. That, for example, there are features of the perceived world that the original hypothesis of water as the basic stuff does not explain might have led to modifications of the latter, so that water hypothesised as the basic stuff comes less and less to resemble water as we find it in rivers and seas; and, indeed, might have eventually been used to explain the properties of ordinary water along with those of other, everyday objects.

As the reducing theories gain in explanatory power, the entities they postulate become further and further detached from the common objects they explain. The other side of the dialectical coin is that, as these reducing theories develop in this way, they may come to affect our view of what it is about the perceived world that needs explaining.

Historically, because of its modest explanatory power, the water hypothesis was not modified in this way but was replaced by other equally mundane candidates. But as long as the basic principles or elements were taken as ordinary things of the world their appearance in explanatory theories was, like the crudities of the modern naked ape theorists, bound to be connected with viciously *ad hoc* arguments if these 'explanations' were even to appear to work.

What was required if this approach was to prove fruitful was the postulation of elements that, although bearing some analogy to the things of the world (and from where else could such an idea initially be got?), remained at some distance from it; and also that the properties attributed to these elements be such that they explain, in some profound way, the characteristics of the world as we perceive it. If such a theory is possible we need not expect it to be entirely correct straight off; but what we might anticipate is that the dialectical relation between it and the level of everyday observation, to which we earlier referred, will be fruitfully established. Such was the atomic theory of Democritus and Leucippus.

This is not the place to go into the question of the original intellectual source of atomism or of its revival in the Renaissance. However, its explanatory power, compared with earlier theories of the fundamental constituents of the world, is evident. Atoms could be said to lie behind the perceived world in the apparently straightforward sense that they were too small to be seen; while, at least at first, they were held to have understandable properties, such as motion, shape, solidity, etc., which were taken from this same perceived world.

These supposed properties of atoms could explain many important features of the world, first in a qualitative way—as when Lucretius explains why water seeps through rock—and then in a precise mathematical way in explanation of such things as the gas laws.

Once established the atomic theory sets going the dialectical process with the result that atoms lose many of their easily understood properties, such as hardness (so that no longer are both the hardness of rock and the malleability of clay, for example, explained in terms of lots of little hard things disposed in certain ways), and take on new and stranger ones.

In this way atomic theory comes to explain more and more things in terms of properties which, although very strange indeed, are saved from the charges of vagueness and magic by their precise mathematical formulation and empirical testability.

CONSTITUENTS OF BLOOD

Blood is a tissue made up of cells in suspension and of plasma containing substances in solution. Forty-five percent of blood volume consists of the formed elements (red cells, white cells, and platelets). Fifty-five percent of blood volume is composed of plasma, which consists mainly of water, some proteins, and the various mineral salts necessary to life.

HOW IS BLOOD MANUFACTURED

Blood is assembled by the body from many different sources. With the exception of injections that may be necessary for treatment, all constituents of blood ultimately come from substances taken by mouth.

The plasma is formed from water, proteins, and salts, all of which are absorbed from the digestive tract. The proteins are modified by the metabolism of the body into the type needed for the human species.The cells of the blood come from diverse sources. The various organs and tissues that collaborate in the manufacture of blood cells are known as the hemopoietic (blood-forming) system.

This includes the bone marrow, spleen, lymph glands, stomach, liver, and the reticuloendothelial system (a network of special tissue cells plus the lining of the blood vessels). The reticulo-endothelial cells are widely distributed throughout the body, but the most important sites are the lining of the bone marrow and spleen.

They are phagocytic; that is, they can ingest micro-organisms and foreign particles of all sorts, thus removing them from the circulating blood.

FUNCTIONS OF THE BLOOD IN THE BODY

Blood, a complex fluid serving many purposes in the human body, preserves the internal environment required for normal activity of the living cells. Following are some of its main functions:

Respiratory. Blood transports oxygen from the lungs to all the body tissues, and simultaneously carries carbon dioxide from these tissues to the lungs, where it is expired. (External respiration.)

Nutritive. Blood conveys food materials from the alimentary canal and food depots to all the tissues of the body.

Excretory. Blood assists in the removal of waste products of metabolism by conveying them to the kidneys, where they are excreted in the urine. The maintenance of the water content of the tissues. Fluid lymph, containing oxygen and food materials, leaves the vascular channels and comes in contact with the tissues. Here it picks up carbon dioxide and other waste products and then flows back into the vascular channels. (Internal respiration.) The complex chemical processes of the body are enabled to take place because of the high dissolving and ionizing property of water in the blood and lymph.

Regulation of body temperature. The body owes its ability to regulate temperature largely to the water content of blood and of the tissue fluid. Heat from the deeper regions of the body is brought to the skin and lungs by the blood stream and dispelled.

Protective and regulatory. Blood and lymph contain complex chemical substances such as antibodies, antitoxins, and lysins—all of which protect the body against injurious substances. The circulating blood also brings hormones from the ductless glands into direct contact with the tissues that respond to such stimuli.

HOW MUCH BLOOD IS IN THE BODY

The average healthy individual has a blood volume equal to about 8 percent of his body weight. Thus the mean blood volume for a person weighing 110 pounds would be 4,400 ml.; for a person weighing 150 pounds, it would be 6,000 ml.; for a person weighing 200 pounds, it would be 7,600 ml. As 1,000 ml. is equal to 2.1 pints, it will be easy for one to estimate the number of pints of blood in his body. The volume for men will be slightly higher than that for women of the same weight because men have an agerage of 7½ percent more red cells than do women.

The amount of blood in the human body has been the subject of careful study. The methods used to calculate blood volume are based on this fact: When a known amount of concentrated substance is introduced into the blood stream, the extent to which the concentrate is diluted is a measure of blood volume. Substances used must be harmless in small quantities and easily identifiable in blood samples. Depending upon the type used, they attach themselves either to the red cell or to a plasma protein. Red cell volume has been measured by tagging the red cells with radioactive molecules; plasma volume has been measured by using a harmless blue dye. Information gained in this way can be used to calculate the whole blood volume.

The blood volume may vary within certain normal limits, depending upon the needs of the body. At rest, the circulating volume is at a minimum. But the blood volume may be increased by several conditions, including hot weather, high altitudes, muscular exercise, emotional excitement, and pregnancy. On the other hand, an abnormal decrease in blood volume may occur during anemia, hemorrhage, loss of plasma due to extensive burns, or by various forms of dehydration.

RED CELLS

Red cells, or erythrocytes, are small, solid, disc-shaped particles whose main function is to transport oxygen from the lungs to the tissues and to transport carbon dioxide from the tissues to the lungs. The red cells contain hemoglobin, a chemical substance that has an amazing affinity for oxygen and carbon dioxide.

Compared to other cells in the body, red cells are incomplete and cannot reproduce themselves because they lose their nuclei shortly before being released from the bone marrow into the blood stream.

However, they are considered to be "alive" in the sense that they are able to perform their functions in the body for as long as 120 days. After that time they wear out and are then removed from circulation and destroyed by the spleen. The various components of protein and iron are salvaged for reuse by the body.

Red cells are created in vast numbers, mainly in the marrow of the short, flat bones of the body. It is estimated that they are worn out and removed from the circulation at the rate of 10 billion red cells an hour. The means by which the bone marrow must produce this same amount in the same period of time is part of the marvelous balance of supply and demand carried on by the normal physiology of the body. Red cells appear fragile in structure under the microscope but are actually very durable. They must be able to withstand a good deal of hammering about while being pumped through the blood vessels, and are able to squeeze through a capillary, which has a smaller diameter than the cell itself. In order to do this, the red cells must fold and bend in a flexible manner and be able to repeat the process many times.

While in the lungs the red cells are exposed to oxygen for only a brief time, but, owing to a chemical balance in the surrounding plasma, they become saturated with oxygen in 1 second. These oxygenated cells are then carried from the lungs to the heart and pumped out into the body. In the tissues, a delicate chemical change takes place in the plasma, and the oxygen is released for use. The carbon dioxide, created by the metabolism of the body cells, is picked up immediately by the red cell in exchange for oxygen and carried to the lungs. Here, another chemical change takes place in the plasma, a change enabling the red cell to release the carbon dioxide and almost instantly pick up more oxygen. This cycle is constantly repeated.

Red cells are very small, measuring about 7.5 microns in diameter and 1.95 microns in thickness. (A micron is about 1/25,000 of an inch.) It would require approximately 3,000 red cells lined up in their widest diameter to cover the space of an inch. Yet, because there are so many red cells in the human body, they would encircle the earth four times if lined up in a beadlike chain. In spite of the smallness of the individual red cell, the surface area of the combined cell mass is very large. In a man of average size, the surface area of the red cells represents a space as large as the surface of a football field.

Because one-fourth of the circulating blood is in the lungs at any one time, a combined cell surface area of 1,200 square yards is available to air. In 1 second, about 370 square yards of red cell tissue surface is exposed to the air, binding the oxygen to the hemoglobin. Only through these unique properties of the red cell, can manwith his relatively large size and weight-obtain enough oxygen to satisfy the needs of his complex system. From the heavy work of the muscles to the delicate functions of the brain, all cells and tissues depend on oxygen for life.

WHITE CELLS

The white cells, or leukocytes, are the protective cells of the blood stream. They are two or three times as large as the red cell and possess a nucleus. In a normal person, there are about 7,000 white cells per cubic millimeter * of blood (or about one white cell to 600 or 700 red cells).

There are several varieties of white cells, and each variety has a different function. When a drop of blood is placed on a glass slide, spread into a thin film, and allowed to dry, it may be stained with a dye so that each type of white cell has a distinctive appearance and colour. White cells may thus be classified into two major groups: those that have easily visible granules in the cytoplasm (the body of the cell exclusive of the nucleus) and those that do not.

The most common type of white cell in the circulating blood has granules and is thus called a granulocyte. This variety comprises 60 to 70 percent of the white cells. The granulocvtes have the ability to migrate, like an amoeba, along the wall of a blood vessel and out of the capillaries into the body tissues, to fight bacteria. Also like an amoeba, they are able to engulf foreign substances. (This property is common to the granulocytes, monocytes, and reticulo-endothelial cells—all sometimes referred to as phagocytes.) When the body is threatened by infection, the granulocytes quickly gather at the point of invasion and set up a first line of defence. As an example, when you prick your finger and the bacteria enter and start to multiply, the body will mobilize many white cells at that point. They call sometimes be seen as a yellow accumulation called pus.

Lymphocytes, a second type of white cell, comprise 25 to 35 percent of the white cell population. These cells—formed by the spleen, lymph nodes, and other lymphoid tissue of the body—do not appear to be so aggressive in combating acute infection, but cluster in the tissues where chronic infection exists. There they form a barrier against the spread of infection to other parts of the body. Lymphocytes may undergo certain chemical changes that are protective in nature, and they may have some function in the development of immunity.

Monocytes, a third type of white cell found in the blood, are larger than the other white cells and relatively few in number. Like the granulocyte, they have the ability to migrate to the site of infection and ingest bacteria. Like a lymphocyte, they tend to wall off the infected area. During the healing process, they are capable of growing into another type of tissue repair cell, strengthening the weakened area.

The disease of leukemia is an abnormal, unrestrained overgrowth of white-cell-forming tissue. In other diseases, there may be an increase or decrease of white cells in the circulating blood. By obtaining a laboratory count of the number of white cells in a cubic millimeter of blood, a doctor has a valuable clue as to the state of health or disease of the body.

PLATELETS

Platelets are small, colorless, cell-like bodies that have an important function in the clotting of blood. They are a source of thromboplastin, a substance that is one of the links in the chain of clot formation. Platelets are liberated into the blood stream as fragments of a large parent cell in the bone marrow. They are smaller than the red cell and fewer in number (one platelet to 10 to 20 red cells). The normal range in platelet counts is quite wide, varying from 200,000 to 400,000 per cubic millimeter of blood.

PLASMA

Plasma is the fluid part of the blood in which the red cells, white cells, and platelets are suspended. It is composed of 92 percent water and 7 percent proteins; the remaining 1 percent consists of fat, carbohydrates, and many mineral salts necessary for life. It also contains hormones, vitamins, and enzymes. Many of the plasma proteins have specific uses in the treatment of disease.

The most well known of these plasma derivatives are gamma globulin, serum albumin, and fibrinogen. Fresh frozen plasma is also prepared for the treatment of certain bleeding problems, especially hemophilia.

Plasma is separated from whole blood by centrifugation; the red cells, being heavier, settle to the bottom of the blood-collecting container. The plasma can then be drawn off into separate containers and processed as fresh frozen plasma or sent to commercial firms for fractionation into specific derivatives.

BLOOD DERIVATIVES

Blood derivatives are those substances that can be separated from whole blood and used for specific purposes. Plasma itself is a component of whole blood. From the plasma there are a number of proteins that can be separated, and these fractions are usually called derivatives. Serum albumin, which in the body maintains plasma volume, is used for shock, low blood protein, and edema (the loss of water from the blood stream into the tissues). Antibodies against disease are concentrated in the gamma globulin.

Fibrinogen is one of the proteins essential for the formation of a blood clot. In addition to serum albumin, gamma globulin, and fibrinogen, a number of other plasma proteins are under study. These may prove valuable additions to the derivatives mentioned above. The red cells left in the bottle after the withdrawal of plasma can also be used to great advantage in certain diseases.

BLOOD CLOT

The clotting of blood is the result of a delicate and complex series of biochemical events, some of which are not fully understood. This mechanism must satisfy two rigorous requirements if life is to be sustained:

- Blood must never clot while it is performing its functions within the intact circulatory system;
- Blood must always clot when any damage occurs that causes a break in the blood vessels. The body must

be able to vary this complex response to bleeding ranging from a slight cut to a massive hemorrhage.

When tissues are damaged, thromboplastin is released from the nearby cells and platelets. This initiates a chain of reactions in which calcium plays an important part. The end result is that the soluble protein (fibrinogen) is converted to a gelatinous solid (fibrin).

Fibrin forms in threadlike strands, making a webbed mesh in which the sticky platelets, red cells, and white cells are trapped, forming a jellylike clot. As the platelets distintegrate, they cause the fibrin web to shrink and have a constricting action on the walls of the torn blood vessel, causing it to clamp down and narrow the tear through which blood is escaping.

This knowledge of clot formation is applied in the collection of blood for transfusions. The calcium, which plays a key role, is inactivated when it is mixed with a citrate solution in the blood container during collection.

ANTIBODY

An antibody is a substance in the plasma that reacts in a highly specific way with a foreign substance (called an antigen) when it is introduced into the body. Although this reaction is usually protective, it may sometimes result in injury to the body.

As an example of its protective function, if the body is invaded by a virulent strain of bacteria, two main lines of defence are set up. The white cells attempt to wall off and engulf the invaders; the bacteria, being a foreign substance, are antigens, and stimulate the reticulo-endothelial cells of the body to produce an antibody that will assist in repelling them. This antibody, contained in the globulin fraction of the plasma, may act on the bacteria in one of several ways. It may dissolve them; it may cause them to clump so that they cannot circulate freely through the blood stream; it may soften their protective membrane and inactivate them so that they will fall easy prey to the phagocytic white cells.

So it is that this type of antibody forms a protective agent against disease. Furthermore, as most of us are exposed to a large number of disease-causing microorganisms during our lifetime, we develop a concentration of various antibodies that may remain in the blood stream for a long time, even after we are well.

If we donate blood, the globulin fraction containing these protective substances can be separated from the rest of the plasma components and concentrated to about 25 times its normal strength. This concentrated substance, called gamma globulin or immune globulin, can then be injected into others to help protect them against certain diseases.

The antigen-antibody reaction also has an important application in giving blood transfusions. Red blood cells, if they are not the proper group and type, may act as antigens in the blood stream of the recipient. In this case, oddly enough, the antibodies in the A, B, AB, and O groups do not have to be developed by the introduction of red cells but exist in the blood stream in the natural state. In contrast to this, the antibodies to the Rh factor do not exist in the natural state and must be developed by a transfusion of incompatible Rh-type cells.

In the case of the ABO groups or the Rh types, the incompatible antigenic red cells are attacked by the circulating antibodies and caused to clump, or agglutinate, in the blood stream. If this reaction is strong enough, the red cells may be broken up, or hemolysed, by the antagonistic antibody. If enough of the transfused, incompatible red cells are clumped or broken up in the recipient's body, they have a damaging effect on the tubules of the kidney, and serious illness or even death may result from the inability of the kidney to excrete waste products. The steps by which these hazards are avoided will be discussed in a later section on laboratory work.

BLOOD GROUPS

Blood groups are inherited patterns into which human blood may be divided for scientific purposes. There are four major blood groups: A, B, AB, and O. * Groups A, B, and O were discovered by Dr. Karl Landsteiner in 1900, and a fourth, AB, was described by another scientist in 1902. Although the percentage of people belonging to each group tends to vary in different countries (as do other inherited characteristics, such as colour of eyes and hair), the following distribution represents the general spread in this country: 45 percent—group O; 40 percent—group A; 10 to 12 percent —group B; 3 to 5 percent—group AB.

About the turn of the century, there was great interest in learning how to use blood transfusions as part of the practice of medicine; but doctors were held back by a forbidding obstacle: while about half the transfusions given to treat illness had a remarkable therapeutic effect, the other half seemed to make the patient much worse and sometimes resulted in death.

In seeking the clue to this riddle, Landsteiner performed a series of tests in the course of which he placed the red cells of one person in suspension in the serum of another person. From this crucial experiment, he noted that in a number of the combinations of red cells and sera, the red cells remained undamaged and freely suspended; in other combinations, the red cells were clumped together in sticky clusters—with resultant damage to the cells.

From this evidence, he deduced that certain combinations of cells and sera were incompatible and that it was these unfortunate combinations that were causing illness and death in some patients following transfusion (transfusion reaction).Further investigation of the factors causing clumping, or agglutination, of the red cells when exposed to incompatible sera revealed that there are two main antigenic substances (agglutinogens) in the red cells and two main antibodies (agglutinins) in the plasma. The antigens in the red cells were named A and B, and the antibodies in the plasma were given names to correspond with the group of red cells that they would agglutinate.

Therefore, the plasma antibody causing group A cells to clump was called anti-A; the plasma antibody causing group B cells to clump was called anti-B.

As noted in the discussion on antibodies, the anti-A and anti-B substances are contained in the globulin protein of the plasma. These basic facts enable us to understand the four simple combinations that are the bases for the major blood groups:

- *Group A blood:* Has A substance in the red cells, and anti-B substance in the plasma.
- *Group B blood:* Has B substance in the red cells, and anti-A substance in the plasma.
- *Group AB blood:* Has both A and B substance in the red cells, and neither anti-A nor anti-B substances in the plasma.
- *Group O blood:* Has neither A nor B substance in the red cells, and has both anti-A and anti-B substances in the plasma.

It will be noted that each blood group is named for the type of antigenic substance that is an inherited characteristic of the red cell.

DETERMINATION OF BLOOD GROUPS IN THE LABORATORY

Blood groups are determined in the laboratory by a simple technique utilizing the known facts about the A and B substances in the red cells, and the anti-A and anti-B substances in the plasma. On the left side of a clean glass slide, the technician places a drop of serum known to contain anti-A substance; on the right, a drop of serum known to contain anti-B substance. He then places a drop of blood (obtained from the person whose group he wishes to determine) into each of the specimens of anti-A and anti-B sera. After a minute or two, the technician can observe one of four possible events take place by which he can determine the blood group:

- If the anti-A serum alone causes the red cells to clump, the blood sample is labeled *group* A.
- If the anti-B serum alone causes the red cells to clump, the blood sample is labeled *group* B.
- If *both* anti-A and anti-B sera cause the red cells to clump, the blood sample is labeled *group* AB.
- If *neither* anti-A nor anti-B serum causes the red cells to clump, the blood sample is labeled *group* O.

The clumping, or absence of it, is clearly visible to the

naked eye in a few moments. Thus, in a well-organized laboratory the technician can determine the groups of a large number of samples in a few hours' time.

For safety, the results of blood group determination are checked in the laboratory in another procedure called proof of blood group. This utilizes samples of known group A and group B red cells, against which samples of the unknown serum are tested. Once again, four observable results may be noted: the group A cells alone may be clumped, the group B cells alone may be clumped, both group A and B cells may be clumped, or no clumping at all will occur.

One can easily work out the blood group that would occur with each of these events. Still another use is made of these facts when the bottle of blood, properly identified as to group, reaches the hospital and is ready to be transfused into the injured or ill patient. Even though the blood is the same group as the patient's, another precaution must be taken to ensure that bloods of donor and recipient are compatible.

This is a laboratory test, very similar to those described above, known as crossmatching. A drop of the donor's red cells is mixed with a drop of the recipient's serum; this is known as the major side of the crossmatching test. The mixture of recipient's cells with donor's serum is known as the minor side of the test. If agglutination occurs on the major side, the blood is incompatible and must never be given because the donor's cells would be agglutinated and hemolyzed in the recipient's blood stream, causing dangerous transfusion reaction.

If agglutination occurs on the minor side, as it will when group O blood is given to A, B, or AB recipients, it does not mean that the blood cannot be given. At the present time, however, it is the practice to give group-specific blood (group A to group A recipients, B to B recipients, and so on) wherever possible and to use group O donors as universal only in emergencies.

HOW ARE BLOOD GROUPS INHERITED

Blood groups are inherited as dominant characteristics according to the Mendelian law: for every inherited characteristic, including blood group, there is a pair of genes, one contributed by each parent. The blood group that a child may inherit is therefore dependent on the combinations resulting from genes contributed by both parents. Owing to intermarriage of people with different blood groups, a variety of combinations can result; but if the parents' blood groups are known, those possible for the children can be predicted.

MEDICOLEGAL IMPLICATIONS OF BLOOD GROUPS

Blood groups are inherited and never change. For this reason, they may have legal importance in identifying bloodstains or determining parentage. But the evidence so provided is useful only in a negative sense.

For example, if a murder suspect had his clothing stained with group A blood, and the victim was of group O, this is evidence that the suspect was not stained by the victim's blood. However, even if the stains proved to be group O, it still would not be conclusive evidence, for the suspect may have had his clothing stained with the blood of any person who was group O.

This logic also applies to paternity cases. The determination of paternity in disputed cases is very complicated. Such cases must be studied by experts in forensic medicine.

THE Rh FACTOR

The Rh factor is an inherited antigenic substance in the red cells (similar in nature to the group A and B substances) that occurs in 85 percent of the white population of this country.

Those who possess this factor are said to be "type Rh positive." Those who do not possess it are "type Rh negative." It occurs in about the same percentage in all blood groups; that is, of 100 group A bloods, approximately 85 will be Rh positive and 15 Rh negative; of 100 group B bloods, approximately 85 will be Rh positive and 15 Rh negative. These same percentages hold for all four blood groups.

The Rh factor was named in 1940 by Landsteiner and Wiener, who were doing work with the red cells of the rhesus monkey. They discovered that the antiserum that agglutinated the cells of the rhesus species also agglutinated 85 percent of the donors they examined; it failed to agglutinate the other 15 percent. The word rhesus was shortened to Rh in discussions of this new factor and thus entered the literature. In 1939 Levine and Stetson reported a severe transfusion reaction in a woman delivered of a dead baby, and discovered that her blood would agglutinate 80 percent of donors who were of her own blood group.

They assumed that the infant had inherited from the father a blood factor that was not present in the mother's blood. Small quantities of the fetal red blood cells had seeped through the placental membranes and caused antibody production in the mother. The brilliant scientific work by which these groups of research workers correlated their separate findings and laid the foundations for the important work on the Rh factor is worthy of additional reading.

SIGNIFICANCE OF THE Rh FACTOR

The Rh factor is of significance in blood transfusion and may be significant in childbirth. The Rh substance can cause a reaction if injected into the blood stream of a person who is Rh negative; that is, his body will have a tendency to break down the Rh-positive cells. This may clog the kidney with the residue of broken-down red cells and, if severe, can result in death.

However, the first tranfusion may not cause this reaction because the recipient may become only sensitized; in other words, he develops anti-bodies—but not enough to produce severe symptoms. However, having been sensitized, he might suffer a severe reaction from subsequent transfusions.

In pregnancy the Rh-positive child of an Rh-positive father may be carried by an Rh-negative mother. In a small percentage of these cases, Rh-positive cells from the infant's blood stream cross the placental barrier membrane and seep into the mothers' circulation.

Because these cells of the infant carry an antigenic substance, Rh, they may evoke in the mother's body an antibody response against the cells. This has little effect on the mother because the amount of the infants' cells in her blood stream is very small compared with the volume of her circulating blood.

However, if the antibodies to the Rh-positive cells generated by the mother are carried back into the blood stream of the infant, they may do a great deal of damage. The concentration of the antibodies in the small body of the infant is so great that it may damage a large percentage of the red cells. Depending on the degree of severity with which this happens, the infant at birth may be merely jaundiced, a condition that may clear up in a few days; he may be quite ill, and to save his life may require an exchange transfusion (removal of baby's blood and simultaneous replacement with Rh-negative blood). In a very few cases, the baby is born dead.

Because the Rh factor has received a great deal of publicity during the past few years, many women fear the possibility of giving birth to an infant affected by hemolytic disease of the newborn (erythroblastosis). Actually, much of this apprehension is unfounded. It should be understood that Rh is a normal inherited characteristic, just as colour of eyes and hair. Again, marriages with an Rh-negative woman and an Rh-positive man may occur in approximately 13 percent of marriages (15% x 85% = 13%).

Most women who are Rh negative do not produce anti-Rh antibodies, and those who do usually have at least two babies that are unaffected. Even in marriages with an Rh-negative woman and an Rh-positive man, only 1 in about 26 has an infant affected, and many of these children only mildly so. With the present knowledge about blood, there is a much better outlook for detecting any Rh incompatability and treating the affected infant.

Chapter 2

Uses of Biochemicals

To deny agriculture the use of these chemical tools would be to jeopardize our agricultural economy and an adequate, well-balanced food supply for the public. Uninhibited insects and plant diseases would largely eliminate the commercial production of such crops as apples, peaches, citrus fruits, tomatoes and potatoes, to mention only a few and would drastically curtail the production of many other major crops.

The claim is often made that without chemicals to protect our orchards and fields from pests, we could not possibly have a modern, productive system of agriculture. There is abundant evidence that our fruit, vegetables and many other staple crops cannot be produced economically, efficiently and in an adequate volume without chemical protection from insects, plant diseases, weeds and other pests, observes George C. Decker, a noted American entomologist.

Although no one would want to jeopardize our agricultural economy, still less deny the American people a well-balanced diet, a great deal of public and professional concern has been aroused by the presence of pesticidal residues in food. An insecticide should be "used as a stiletto rather than a scythe," A. W. A. Brown writes. "Ideally it should be employed to punch a particular hole in the food chain, which, if it cannot be filled in by [other species in the] neighboring elements of the biota, at least should be done at such a time of season that it does not seriously disturb the biota."

In practice, the trend has gone in the opposite direction. Insecticides are not only becoming more lethal; they are being used with less and less discrimination. Aside from the ecological implications of this trend, the growing toxicity and variety of insecticidal residues in food have become one of the major health problems of our time.

The large-scale application of insecticides to field crops did not get under way until the 1860's, when a large number of American farmers began to use paris green (copper acetoarsenite) against the Colorado potato beetle. The effectiveness of this insecticide encouraged chemists to develop a variety of other formulas. Most of the new insecticides were inorganic compounds—preparations of arsenic, mercury and sulfur—which normally remained on the crop after harvesting. The residues ranged in amount from mere traces to quantities large enough to be hazardous to the consumer.

As early as 1919 the Boston Board of Health condemned shipments of western pears because of excessive arsenic residue and six years later British health authorities raised strong objections to the amount of arsenic found on imported American apples. Finally, federal officials placed limits on the amount of arsenic that could be allowed to remain on commercial apples and pears. Ironically, for a brief period the original limits, or tolerances, for food sold to domestic consumers were two and a half times higher than those established for apples exported to Europe.

Although the prewar era of inorganic insecticides has been described as pristine in comparison with the present era of organic preparations, the situation was far from ideal. In the 1930's one of the most widely used inorganic compounds, lead arsenate, combined a suspected carcinogen (arsenic) with a cumulative poison (lead). Calcium arsenate seriously damaged the soil, especially soils low in humus. Many areas of Washington and South Carolina were rendered unproductive as a result of being dusted with large amounts of arsenicals.

Harmful as they are, however, prior to 1945 inorganic insecticides were employed in relatively modest quantities and in a limited variety. The most lethal preparations were used primarily for agricultural purposes and the greater part of the residue could be removed from the surface of the crops by careful washing. After World War II, the insecticide problem acquired entirely new dimensions. The total quantity of insecticides used prior to the war was but a fraction of the amount employed in the mid-1950's. Annual domestic sales of pesticides (primarily insecticides) increased more than sevenfold between 1940 and 1956. Inorganic insecticides, though still in use, declined markedly in importance; they were supplanted for the most part by a whole new group of organic preparations.

Of these the best known is DDT, a chlorinated hydrocarbon; it was synthesized as early as 1874, but it remained a -known chemical compound until the late 1930's, when the Swiss chemist Paul Müller discovered its insecticidal properties. The American armed forces began using it in 1943, a year after the first small quantities had been shipped to the United States for experimental purposes. While the war was still in progress, A. P. Dupire in France and F. J. Thomas in England began to work with benzene hexachloride (BHC), another chlorinated hydrocarbon that had been synthesized in the last century.

The research of Thomas and his colleagues led to the discovery of the insecticidal properties of lindane, a form of BHC. Significant work was also being done in Germany during this period. Shortly before the war, Gerhard Schrader, while engaged in research on poison gases, discovered a series of extremely lethal organic phosphates, from which parathion, malathion and other now familiar preparations were developed. The list continued to grow until an impressive number of chlorinated hydrocarbons and organic phosphates were available for use on farms and in gardens, warehouses, stores and homes.

In retrospect, the carelessness with which some of these insecticides were delivered to the public is incredible. The chemical behaviour of the chlorinated hydrocarbons inside the human body is still unknown, although it has been determined that both the chlorinated hydrocarbons and the organic phosphates are powerful nerve toxicants. The organic phosphates are known to inactivate the enzyme cholinesterase, which participates in the chemical control of nerve impulses.

Solutions of organic insecticides are absorbed through the skin as well as the lungs and digestive tract. The skin does not offer an effective barrier to the entry of these toxicants when they are used in liquid form.

The younger an organism, the more easily it is poisoned. Both groups of insecticides produce effects in warm-blooded animals very similar to those they produce in insects. Farm animals poisoned by DDT and other chlorinated hydrocarbons first become restless and excitable; the next stage is characterized by twitching, spasms and, finally, convulsions. Animals poisoned by the organic phosphates usually die of respiratory failure, but in these cases, too, convulsions have been known to occur.

Fortunately, DDT, the first of the new organic insecticides to be widely used, is one of the milder chlorinated hydrocarbons. The compound began to reach American farms in 1945. As test samples of DDT had arrived in the United States only three years earlier, American research on the compound up to that time had been necessarily limited. Some of its more disquieting features were still under study when many Americans began using it with reckless abandon as a "magic bullet" against insect pests.

At this stage, very little was known about the way the human body reacts to DDT, the possibility of chronic toxicity, the existence of the indirect routes by which DDT enters the human body and DDT's stability under varying agricultural conditions. By 1950 when a number of these questions had been answered, it had become evident that although the insecticide is certainly a "magic bullet," the target toward which it is traveling is very much in doubt.

It was found that DDT tends to concentrate heavily in fat. A relatively low dietary intake of the insecticide is magnified from six to twenty-eight times in the fat deposits of the body. According to A. J. Lehman, of the F.D.A., the continual ingestion of as little as 5 parts per million (ppm) of DDT in the diet produces "definite though minimal liver damage" in rats. Such an intake is not likely to be uncommon in the United States. The F.D.A. has placed a residue tolerance of 7 parts per million of DDT on nearly every fruit, vegetable and portion of animal fat that reaches the American dinner table. But the agency does not know how many farmers are honoring these tolerances. The reputation of DDT suffered its first blow only two years after the end of the war, As early as 1947, Italian entomologists noted that DDT-resistant flies were appearing in many parts of Italy. Similar reports began to come from California and other areas in the United States. By 1950, American surveys had shown that previously susceptible strains of flies "were becoming resistant to DDT over a period of 2 or 3 years."

Entomologists and public health workers were still hopeful that the malaria-carrying mosquito would continue to succumb to DDT, but this insect also developed DDT-resistant strains. As the number of resistant insects began to multiply, attention shifted to more lethal preparations, such as chlordane, dieldrin and the organic phosphates.

CHLORDANE AND THE DIELDRIN-ALDRIN GROUP

Chlordane and the dieldrin-aldrin group of preparations are highly toxic chlorinated hydrocarbons. A portion of the testimony and data that the Delaney Committee received on chlordane is summarized in its report as follows: "The Director of the Division of Pharmacology of the Food and Drug Administration testified that from a chronic viewpoint chlordane is four to five times more toxic than DDT and that he would hesitate to eat food that had any residue on it whatever.

He stated that chlordane has no place in the food industry where even the remotest opportunity for contamination exists and that it should not be used even as a household spray or in floor waxes." Later experimental work showed that chlordane, when fed to rats, has about twice the chronic toxicity of DDT; the insecticide produces detectable liver damage when ingested in amounts as low as 21/2 parts per million.

Dieldrin and aldrin, according to Lehman, create a danger of chronic toxicity that "may be similar" to that posed by chlordane. These insecticides have the "rather rare property of being more toxic by skin absorption than by oral ingestion, the ratio being about 10:1. To state this another way, dieldrin and aldrin are approximately 10 times more poisonous when applied to the skin than when swallowed. Neither of these compounds is irritating to the skin in low concentrations so that the individual has no warning of skin contact."

The F.D.A. permits residues of chlordane, dieldrin and aldrin on nearly all the vegetables and fruits sold to American consumers. In view of current spraying practices, it is not unusual for foods to accumulate residues of several chlorinated hydrocarbons. The fact that the various insecticides have different chemical properties is small reason for comfort. "Studies in our laboratories have shown that when low concentrations of several insecticides are fed to rats the effect is additive," Lehman writes.

"For example, when 2 ppm each of DDT, chlordane, lindane, toxaphene and methoxychlor, making a total of 10 ppm of chlorinated hydrocarbons, are added to the diet of rats, liver damage results. With the possible exception of chlordane, none of the insecticides listed, when fed individually at 2 ppm, would injure the liver, but in their combination their effect is additive." Where do we stand today in our chemical war against the insect world?

Although partial control has been achieved over some carriers of disease and other pests are no longer the nuisance they once were, the basic problems of insect infestation remain unsolved. Herbert F. Schoof observes that "man's insect enemies are adapting to successful coexistence with pesticides almost as rapidly as new technics and toxicants are produced." We are paying a heavy price for our questionable gains against agricultural pests. A larger quantity of insecticides is required almost every year.

The production of DDT in the United States nearly doubled between 1950, when the compound was no longer a novelty and 1958. Nearly 100 million pounds of aldrin, chlordane, dieldrin, endrin, heptachlor and toxaphene—chemicals virtually unknown to farmers in the 1940's—were produced in 1958. Approximately 3 billion pounds of pesticides, including herbicides, fungicides and fumigants, are produced annually in the United States. "The pesticide industry is growing by leaps and bounds and entomologists predict and chemical manufacturers hope for, a fourfold expansion in use of pesticides during the next ten to fifteen years," observes Clarence Cottam, of the Welder Wildlife Foundation.

"Today, well over 12,500 brandname formulations and more than 200 basic control compounds are on the market. Most of the currently used pesticides were unknown even ten years ago. Furthermore and contrary to the public interest, most new pesticides are decidedly more toxic, generally more stable and less specific in effect than those of but a few years back."

The postwar era also witnessed the introduction of highly potent chemical agents for increasing the weight of livestock. In 1947 the F.D.A. approved the use of stilbestrol pellets for "tenderizing" the meat of poultry. Stilbestrol, a synthetically prepared drug, causes bodily changes that are identical with those produced by certain female sex hormones (estrogens). The drug was approved for use with the understanding that poultrymen would implant a single pellet in the upper region of the chicken's neck about a month or two before marketing,

Presumably this period was sufficient for the complete absorption of the chemical, so that no residues would be left in the flesh of the bird. Male birds treated with stilbestrol undergo radical physiological changes; their reproductive organs, combs and wattles shrivel and they lose all inclination to crow or fight. The drug tends to cause substantial increases in weight in both sexes, producing plump birds with skin that is noticeably smooth in texture.

Careful investigation has shown that stilbestrol increases the over-all fat content of chickens by 40 per cent. In the hearings before the Delaney Committee, Robert K. Enders, professor of zoology at Swarthmore College, expressed his agreement "with those endocrinologists who say that the use of the drug to fatten poultry is an economic fraud. Chicken feed is not saved; it is merely turned into fat instead of protein.

Fat is abundant in the American diet so more is undesirable. Protein is what one wants from poultry. By their own admission it is the improvement in appearance and increase in fat that makes it more profitable to the poultryman to use the drug. This fat is of very doubtful value and is in no way the dietary equal of the protein that the consumer thinks he is paying for."

By the late 1950's approximately 100 million hormone-treated fowl were being marketed annually to American consumers. In a number of limited surveys, the F.D.A. found that stilbestrol pellets often remained in the necks of poultry until after the birds had reached wholesale outlets; the pellets were not completely absorbed. Stilbestrol was detected in the edible tissues of hormone-treated fowl even after the heads and necks had been removed. Finally, in December 1959, the F.D.A. announced that, on its recommendation, the poultry industry had agreed to discontinue the use of the synthetic estrogen.

It is doubtful whether either the F.D.A. or the poultry industry had very much choice. Stilbestrol is suspected of being a cancer-promoting drug and recent federal legislation specifically prohibits the presence of carcinogenic additives in food. Pellets of stilbestrol are still implanted in the ears of cattle, however and the drug is widely used as a supplement to the feed of steers and sheep. The official opinion today is that no residues of the drug will remain in large animals if breeders obey F.D.A. regulations, but the evidence to support this opinion is open to question.

The use of synthetic estrogens in livestock management was soon followed by the addition of antibiotics to feed, a technique that dramatically accelerates the growth of domestic animals. Although there is no evidence that antibiotics affect the nutritive quality of meat, research conducted by the F.D.A. showed that chlortetracycline and penicillin could be detected in the blood, tissues and eggs of chickens given feed containing as little as 50 parts per million of the drugs.

This quantity does not exceed the limits established by the F.D.A. for antibiotics in feed. According to William A.

Randall, of the F.D.A.'s Division of Antibiotics, the "small amounts found in chicken tissue were destroyed following frying. Furthermore, when the antibiotic-containing feed was withdrawn and normal feeding begun the antibiotic disappeared from the tissue or eggs within a few days. After hard-boiling eggs containing antibiotics, no activity could be detected in the great majority of those tested."

But we should place the emphasis where it properly belongs. Antibiotics could be detected even in eggs that were hard-boiled and residues were found in the tissues and eggs of chickens after normal feeding was restored. Accordingly, current F.D.A. regulations prohibit the use of antibiotics in the feed of egg-laying hens and require that, in the case of poultry destined for human consumption, normal feeding be restored four days prior to slaughtering. "It may well be in the public interest to reserve this valuable class of drugs for the purpose for which they were originally developed," Randall concludes, "that is, the prevention and treatment of disease."

This conclusion is not without merit. Many farmers pay very little attention to the F.D.A.'s regulations and nearly all farmers ignore them at one time or another. The situation is similar to that faced by the police in trying to keep motorists from speeding; almost everyone occasionally exceeds the speed limit. But in the case of the F.D.A. regulations, the violator's chances of being caught are infinitesimal in comparison with those of the speeder. Although its inspection programme is totally inadequate to cope with the problem of chemicals in food, the F.D.A. has already discovered violations of its regulations on feed supplemented with arsenic compounds, another widely used promoter of growth.

"A very preliminary survey made some time ago by the Food and Drug Administration indicated that some poultry raisers are not withholding arsenic-containing feeds from their flocks 5 days before slaughter," complained Arthur S. Flemming, former Secretary of Health, Education and Welfare, "and we have information that in some parts of the country, hog raisers maintain their animals on arsenic-containing feed within the 5-day period that the arsenic is supposed to be withheld. It is evident that such disregard for safe use could lead to added arsenic residues in poultry and pork on the retail market and thus could frustrate the public health safeguards upon which the original approvals were granted."

ANTIBIOTICS ARE DANGEROUS FOOD ADDITIVES.

The sensitivity of many persons to penicillin has reached a point where a therapeutic dose of the drug can be expected to claim their lives. Such individuals are increasing in number with every passing year. Cases have already been reported in which penicillin used to treat infections in dairy cattle entered the milk supply in sufficient quantities to cause severe reactions in human consumers. According to Murray C. Zimmerman, of the University of Southern California, data on milk contamination suggest that a person who drinks two quarts of milk every day "could ingest over 1,000 units of penicillin daily.

The fact that penicillin 'naturally' present in milk has not previously been proved to cause allergic reactions is outweighed by the fact that *this is almost a billion times more penicillin than has previously been shown necessary to provoke reactions*. Coleman and Siegel reported a patient with a reaction on passive transfer to 0.00001 units of penicillin. Bierlein's patient went into shock when skin-tested with 0.000003 units of penicillin." Zimmerman then proceeds to discuss detailed case histories of four patients whose reaction to penicillin could be traced with reasonable certainty to the consumption of dairy products.

It is quite conceivable that the extensive use of antibiotics in agriculture will slowly turn our domestic animals into a major reservoir of antibiotic-resistant bacteria. This represents a problem fraught with incalculable dangers to public health. Feed supplemented with tetracycline, for example, has already produced resistant strains of *Bacterium coli* and *Clostridium perfigens*. Both bacteria typically inhabit man's alimentary tract, but under certain circumstances they behave as diseasecausing organisms.

"Examination of pigs and poultry kept on many different

farms showed that the *B. coli* in the feces of tetracycline-fed animals were predominantly tetracycline-resistant," observes H. William Smith, of the Animal Health Trust in Sussex, England, "while those in the feces of animals on farms where tetracycline feeding was not practiced were predominantly tetracycline-sensitive. In some herds in which tetracycline feeding was just being introduced it was possible to trace the changes of the *B. coli* fecal flora from tetracyclinesensitive to tetracycline-resistant." Smith notes that similar changes could be traced in the case of *Clostridium perfigens*.

The use of shotgun methods to treat or forestall udder infections in cattle with penicillin has almost certainly increased the propagation of resistant strains of *Staphylococcus aureus,* the principal culprit in the common "staph infection" and the direct cause of grave, often fatal cases of pneumonia. Smith notes that the proportion of penicillin-resistant *Staphylococcus aureus* cultures isolated from herd samples of milk in Britain "had increased from 11% in 1954 to 47% in 1956." Two bacteriologists of the Canadian Food and Drug Directorate, F. S. Thatcher and W. Simon, in analyzing cheese that had been purchased in retail outlets, found substantially more penicillin-resistant staphylococci "than would be expected in a normal population not exposed to antibiotic therapy. With regard to the streptococci recovered from cheese our results show a close approximation to the degree of resistance to penicillin found among comparable species isolated from hospital patients."

Thatcher and Simon conclude their report with the warning that "where penicillin or other antibiotics are used with dairy cattle, the survival of resistant organisms may lead to widespread distribution of resistant strains into the homes of the general populace, since staphylococci and streptococci, often in large numbers, are of common occurrence in cheese. Foods so contaminated may well contribute to the severity of the problem arising from infections in man with resistant strains without the patients having received prior antibiotic therapy or without having been exposed to endemic infections within hospitals."

But in the United States this is no longer merely a potential hazard. Already there have been two localized epidemics of "staph infections" that have been attributed to contact with cattle. During 1956 and 1957 about 36 per cent of the senior students and 26 per cent of the faculty at the Veterinary School of the University of Pennsylvania contracted "staph infections." In 1958 another epidemic of the infections occurred among 44 per cent of the senior students at the Colorado State University College of Veterinary Medicine. There is good reason to believe that, in both cases, animals handled by the students had become a reservoir of antibiotic-resistant staphylococci.

The current trends in agriculture reflect a shocking indifference to the long-term health of our population. By recklessly overcommitting ourselves to the use of chemicals in the growing of our food, we have turned highly potent toxicants and drugs, which we would be justified in using only on a limited scale or in exceptional circumstances, into the foundations of our agriculture. These foundations are extremely unstable and the edifice we are trying to construct may prove to be a trap rather than a habitable dwelling. Insecticides commonly aggravate the very problems they are meant to solve. The ecological difficulties they create lead to the use of increasingly toxic preparations, until a point is reached where the insecticides threaten to become more dangerous in the long run to the human species than to the insects for which they are intended. The need to rescue a crop from agricultural mismanagement gains priority over the health and wellbeing of the consumer. The use of drugs to fatten and accelerate the growth of domestic animals is playing havoc with the physiology of our livestock.

The damage these drugs inflict on poultry and cattle necessitates the use of additional drugs, which, in turn, contaminate man's food and reduce the value of these agents in combating disease. It would seem to be a form of ecological retribution that the very forces man has summoned against the living world around him are being redirected by a remorseless logic against the human organism itself.

THE CONSUMER AND COMMERCIAL FOODS

With the rise of an urbanized society, the production of food becomes a complex industrial operation. In contrast with earlier times, when very few changes were made in the appearance or the constituents of food, much of the food consumed in the United States is highly processed. Allen B. Paul, of the Brookings Institution and Lorenzo B. Mann, of the Farmer Cooperative Service, have summed up the change as follows: "Our grandparents used for baking about four-fifths of the flour milled in this country.

They churned almost all the butter Americans ate. They killed and prepared much of the meat eaten. They made their own soups, sausage, salad dressing, clothing and countless other items. Such tasks, which a generation ago were part of farm and home life, have been taken over by commercial factories, 85,000 of them." Of the 85,000 factories cited by Paul and Mann, nearly half were food-processing plants.

It is hardly necessary to emphasize the fact that advances in food technology and methods of transportation have given us many advantages. A large number of seasonal fruits and vegetables are now available throughout the year in canned and frozen form. Foods are generally cleaner than they were a half century ago. Processed foods have enormously reduced the homemaker's burden. Many vegetables, fruits and meats are shredded, eviscerated, mixed, creamed, cooked, baked, or fried before they reach the kitchen. The homemaker often needs little more than a can opener, a few utensils and a stove to serve meals that required elaborate preparation and hours of work a generation or two ago.

But the shift from home-prepared to processed foods has not been an absolute gain. Although very little was known about human nutrition a few generations ago, our forebears could make use of what knowledge they had in growing and preparing their own foods. Today the situation is different. A great deal has been learned about the nutrients required for human health, but the cultivation, storage, refining and, in large part, the preparation of food is no longer controlled by the consumer.

These tasks have been taken over by the food industry. In the absence of a nation-wide food-grading programme that would compel processors to disclose the nutritive values of food items, the housewife's attempt to plan a healthful diet depends to a great extent upon chance. The nutritive quality of a given food may vary enormously. The variation may be due to the type of seed that the farmer plants, the methods employed 91 in food 91 cultivation and food processing, or the length of time the food is stored.

For example: Nutritionists, in analyzing thirty-one strains of cabbage, found that the ascorbic-acid, or vitamin-C, content varied as much as 35 per cent, while the amount of carotene in different varieties of sweet potatoes has been found to range from zero to 7.2 milligrams per cent. A threefold difference was discovered in the niacin content of forty-six strains of sweet corn. Large variations in nutritive value have been found among different varieties of apples, peas, wheat, onions and many other crops.

The fact that a given variety attains great size does not necessarily mean that it is nutritious. The very opposite may be true. Small cabbages, tomatoes and onions, for example, contain more vitamin C than those of greater size. Size, appearance, colour and texture are often very poor criteria for judging the nutritive content of a food.

Generally, plant breeders are interested in developing varieties of vegetables and fruits that have an attractive appearance, yield more bushels to the acre, are more resistant to disease and withstand storage and shipment. "It is now realized that in the development of these commercially improved varieties by genetic selection the nutrient content is often decreased," notes Robert S. Harris, of the Massachusetts Institute of Technology. "If the plant breeder were to collaborate with a food analysis laboratory, he could develop commercially improved varieties which are also superior in nutrient content."

Foods may lose much of their nutritive value when they are heated prior to canning or freezing. If the processor follows a good blanching procedure, vegetables will retain most of their ascorbic acid and vitamin B. But if the procedure is poor, the losses may be enormous. Losses may run as high as 30 per cent of the riboflavin in green beans, 37 per cent of the naicin in spinach, 40 per cent of the ascorbic acid in peas and 64 per cent of the thiamine in lima beans. "Blanching procedures are not standardized," Mildred M. Boggs and Clyde L. Rasmussen observe. "Some processors have a leaching loss of 5 to 10 per cent of each water-soluble vitamin or other constituents. Others have losses of 40 to 50 per cent—or even higher with the same food."

A substantial loss of nutrients takes place when wheat is refined into flour for white bread. "Even with 4 per cent milk solids in the white enriched bread," Bogert notes, "it is still not quite up to the content of whole wheat bread in protein, calcium, iron, thiamine and niacin, while the whole grain breads contain other B complex vitamins not added in the enrichment of white flour." If a product is stored too long, it may lose a substantial amount of thiamine. "Pork luncheon," for example, may lose 20 per cent of its thiamine in the can while standing on the retailer's shelf for three months. After six months the loss may be as high as 30 per cent.

Many of these problems could be solved by establishing a compulsory system of grade labeling for processed foods based on their nutritive constituents. The grade markings that currently appear on labels are often meaningless from a nutritional point of view. For the most part, they indicate the food's size, appearance and taste, not its nutritive value. But the food industry has been adamantly opposed to any kind of compulsory grade labeling. Jesse V. Coles, professor of home economics at the University of California, observes that "practically all, if not all, manufacturers, processors and distributors are opposed to compulsory grade labeling."

In the thirties the food industry defeated attempts to provide for grade labeling in the N.R.A. fair competition codes; it was bitterly opposed to the inclusion of standards of food quality in the food and drug legislation enacted by Congress in 1938; and it successfully blocked efforts to link price regulations with food quality in wartime anti-inflationary laws.

In the absence of a meaningful system of grade labeling, the homemaker is surrounded by nutritional anarchy. Her choice of a product is often governed by its appearance; if it is packaged, she is likely to choose a well-advertised brand or one that is attractively wrapped. Her difficulties are complicated by the fact that many processors colour the food she sees and use chemicals to soften the refined products she touches.

Preservatives are often added to foods to permit longer storage and synthetic flavoring matter is put into many processed foods to enhance their taste. And even in those cases in which appearance, taste and softness are true indications of superior nutritional quality, the processor frequently uses a large variety of chemicals to make an inferior, stale, or overly refined product seem high in quality.

Aside from the deception they practice on the consumer, many of the chemicals added to foods are known to be active toxicants. Presumably they are "safe under the conditions of their intended use," to borrow a phrase formulated by the Food and Drug Administration, but the word "safe" does not mean that the chemicals do no damage. In view of the fact that they are harmful even in relatively small quantities, it is likely that toxicants in foods inflict a certain amount of damage on nearly all animal organisms, including man.

Another group of food additives is enveloped in an air of biochemical mystery; very little is known about how they react during animal metabolism. Many additives are chemically unique or "exotic." Until they were developed by chemists, the compounds never appeared in foods and their long-range effects on the human body have yet to be determined.

The overwhelming majority of people in the United States derive their nourishment from commercial foods. If a harmful chemical finds its way into a popular food, many individuals are likely to be affected. These individuals differ in age, state of health, genetic endowment, environmental background and susceptibility to the toxicity of the chemical. Some individuals may be more vulnerable to the toxicant than others, although they may never show any symptoms of acute illness.

The research devoted to chemicals in foods seldom goes beyond the effects they produce in laboratory animals. A chemical additive is tested on rats and dogs for about two years. If the chemical causes no apparent damage to tissues and organs, it may be used in food. If it is obviously toxic "under the conditions of its intended use," it will be rejected. But laboratory animals are not men. Their physiology is different from that of human beings and they are unable to communicate important reactions that may well escape the notice of the researcher. Although experiments on rats, dogs and even monkeys are useful in forewarning man against harmful substances, they do not supply conclusive evidence that chemicals which cause no apparent damage to animals are harmless to man.

Evidence that a chemical is harmful may well appear years after it has been permitted in foods. The use of additives regarded as safe only a decade or two ago has been forbidden because later, more exacting studies showed that these chemicals have toxic properties. According to *Consumer Reports*, coal-tar dyes were certified by the Food and Drug Administration for use as food colors "on the basis of hasty, old-fashioned and otherwise inadequate testing and recent advances in research methodology have led to the conclusion that the safety of most of them is unknown.

Since 1945, when the FDA began to apply modern methods of study and research to certifiable dyes, 15 food dyes have been re-examined for toxic, carcinogenic or allergenic properties. Only one of these, Yellow No. 5(used to colour candies, icings and pie-fillings, for example) has been conclusively shown to be harmless. Last year, Orange No. 1, Orange No. 2 and Red No. 32 were decertified as too toxic for use in foods. The FDA announcement said that 'while manufacturers will no longer label and sell these three colors for use, they may label and sell them for external drug and cosmetic use.'

Orange No. 1 had been widely used in candy, cakes, cookies, carbonated beverages, desserts and such meat products as frankfurters. Orange No. 2 and Red No. 32 were used to colour the outer skin of oranges and during the Christmas season last year, some 150 children were made ill in California as a result of eating popcorn colored with Red No. 32." Whether a food additive is classified as harmful or not depends not only upon laboratory techniques but on the standards of the researcher. Every investigator is confronted with the following questions: How thoroughly should the chemical be studied? On how many species? What constitutes reasonable evidence that a chemical additive is harmless to man?

In hearings before the Delaney Committee more than a decade ago, many conflicting answers were given to these questions. Some eminent researchers advanced the view that tests should be conducted through the whole life span of many animal species, while others seemed to be satisfied with limited experiments on rats. The generally accepted criteria for evaluating the toxicity of food additives represent an unstable compromise between widely divergent scientific views and are subject to serious challenge.

While disputes over the adequacy of current testing standards continue, the problem of chemical additives in food becomes greater with every passing year. Chemicals are used in nearly every phase of the production, storage and merchandising of food. It is difficult to find a single commercial food that has not been exposed to a chemical compound at some stage of cultivation or processing. Many foods accumulate a large number of residues and additives as they travel from the farm through the processing plant to the retail outlet.

Grain, for example, is dusted repeatedly with insecticides and fungicides and then treated with fumigants during storage. At the processing stage, chemicals are added to facilitate handling and to improve the consistency of the flour. The flour is bleached and treated with preservatives. The final product may be colored and wrapped in chemically treated paper that sometimes transfers undesirable substances to the food. Are all of these chemicals indispensable to an ample supply of food? Do any of them constitute hazards to public health? At a time when the number of chemicals in our food is reaching staggering proportions, these questions require considered replies.

CHEMICALS IN FOOD PROCESSING

The same over commitment to potentially risky methods is found in other phases of food production. Complex problems of storage, cleansing, handling, refining, cooking, mixing, heating and packaging are often solved by the use of chemical additives. A half century ago the yellowish tint in freshly milled flour was removed slowly by aging; today it is removed rapidly by oxidizing agents.

The miller no longer has to store flour and wait for it to mature. Emulsifiers are employed, partly to "improve" the texture of a product, partly to speed up processing. Their "functional advantages" in ice cream, for example, are that the "mix can be whipped faster, the product from the freezer is dryer and the mix has less tendency to 'churn' in the freezer."

Chemical antioxidants are used to help prevent oils and fats from becoming rancid, a problem that becomes especially pronounced when natural antioxidants are removed during large-scale processing. Chemical preservatives are added to food to permit longer storage in wholesale depots and to extend the "shelf life" of a product in retail outlets. In time chemicals are turned from mere adjuncts of food production into "technological necessities." Their use may result in new machines and facilities, the abandonment of old processing methods and a broad reorientation of technology to meet the new chemical requirements.

The fact that a chemical is technologically useful does not mean that it is biologically desirable. The two phrases are occasionally used interchangeably, as though they were synonymous. Some of the chemicals most widely used in the production of food have been found to be harmful to animals in laboratory tests. Nitrogen trichloride, or agene, was employed for decades as a bleaching and maturing agent in flour before it was discovered, in 1946, that the compound produces "running fits," or hysteria, in dogs.

The use of agene in the bread industry has since been discontinued. Although there is no evidence that agenized flour causes nerve disorders in man, no one can say with certainty that cereal products treated with agene are completely harmless to consumers, especially if eaten day after day for twenty years or more. The toxicant in agenized flour "produces nervous disturbances in monkeys, but not epilepsy in man," observed the late Anton Carlson, one of the most distinguished physiologists of his day. "That is not sufficient for me, because there are many other types of nervous disturbances that may be aggravated or introduced by this chemical."

Another group of chemicals, notably certain types of polyoxyethylene derivatives, were widely used in the United States as emulsifiers before they were banned as potential hazards to consumers. Former F.D.A. Commissioner Charles W. Crawford is reported to have described the agents as "good paint removers." According to data furnished to the Delaney Committee, rats that had been fed two commercial preparations of polyoxyethylene-lb/> derived emulsifiers showed blood in their feces and developed a variety of kidney, abdominal, cardiac and liver abnormalities.

These preparations were used primarily as bread softeners but also found their way into other foods. The polyoxyethylene derivatives constitute a highly suspect family of emulsifying agents, yet some of them are still added to cake mixes, cake icing, frozen desserts, ice cream, candies, dill pickles, fruit and vegetable coatings, vitamin preparations and many food-packaging materials.

Similar examples can be culled from other groups of chemical aids. In general, the toxic effects of many commonly used compounds are difficult to determine, but individually and in groups they have aroused suspicion and concern.

"The fortification of oils and fats against oxidation and rancidity is another field in which there are serious questions of possible toxicity," warns F. J. Schlink, technical director of Consumers' Research and editor of *Consumer Bulletin*. "The natural antioxidants which delay rancidity are lost during the factory processing of refined oils and fats; then the attempt is made to restore the antioxidation qualities by addition of fat-stabilizing substances. Most of the materials that have antioxidation properties are known to be toxic to some degree."

The term "technological aid" has been used loosely to include food additives that could be dispensed with entirely if a manufacturer improved his processing methods and the quality of his product. The question of whether a chemical, instead of being legitimately employed to furnish distant consumers with a perishable product, was being used rather as a device for masking inferior ingredients and unsanitary methods arose in sharp form early in the century, when sodium benzoate was employed extensively as a preservative in catsup. Harvey Wiley, the first administrator of the federal food and drug law, regarded sodium benzoate as a hazard to consumers.

Supported by data from feeding experiments performed with human beings, Wiley claimed that the preservative causes injury to the digestive system. He was convinced that catsup need not deteriorate readily if clean facilities and proper ingredients are used in preparing it. Challenged on this point by Charles F. Loudon, a canner in Terre Haute, Indiana, Wiley appointed a bacteriologist, Arvill W. Bitting, to work at the Loudon factory and prepare a formula for a stable catsup preparation without artificial preservatives. Bitting was successful.

Moreover, after inspecting numerous canning factories and examining many brands of catsup, Bitting concluded that manufacturers who needed sodium benzoate often used inferior products and were careless about sanitation.

"The whole spirit and tradition of the pure-food law is against the use of preservatives and substitutes," declared Mrs. Harvey Wiley more than forty years after the Loudon episode. "The intent of the law is to encourage the manufacture of high-grade ingredients, not to try to hide inferiority and cheapness by the use of chemicals and preservatives." If this is so, the spirit of the law has been ignored for decades.

The most up-to-date surveys of chemical additives in foods show that preservatives are added to cheese, margarine, cereal products, jams, jelly and many other processed foods. Uncooked poultry is dipped in solutions of antibiotics. Aside from the damage to public health that may arise from the use of questionable chemicals as preservatives, the longer "shelf life" acquired by some of these foods may cause a loss of valuable nutrients.

CHEMICAL ADDITIVES IN FOODS

Many chemical additives in foods perform no technological or nutritional function whatever. They cannot be regarded as materially useful by any stretch of the imagination. These chemicals enhance the colour of a food or make certain products feel soft and newly prepared; in some instances they have been used to replace costly but valuable nutrients with inferior ones. In the least objectionable cases, an additive will deceive the consumer without impairing his health.

The application of an innocuous vegetable dye to some foods often leads the consumer to believe that he is acquiring a better, more wholesome, or tastier product than is actually the case. In the most objectionable cases, a chemical is toxic to a greater or lesser degree. This type of additive not only serves to change the appearance or conceal the nutritive deficiency of a food; it exposes the consumer to a certain amount of damage.

Nitrates and nitrites, for example, are used to impart a pink colour to certain brands of processed meat. Ostensibly, this is their sole function. The addition of these chemicals to meat, especially frankfurters and hamburger meat, would be objectionable if only because of the deception which makes it difficult for the housewife to distinguish between high-quality and low-quality products. But this is not the only deception perpetrated on the consumer. Used together with salt, nitrite compounds definitely extend the "shelf life" of processed meat products.

When nitrites are ingested, they react with hemoglobin

in the blood to form methemoglobin and, like carbon monoxide, reduce the hemoglobin's capacity to carry oxygen. An individual who consumes three to four ounces of processed meat containing 200 parts per million of sodium nitrite (a permissible residue) ingests enough of the compound to convert from 1.4 to 5.7 per cent of his hemoglobin to methemoglobin. Ordinarily, this percentage is insignificant. But if the same individual is a heavy smoker and lives in an air-polluted community, the cumulative loss in oxygen-carrying capacity produced by the nitrites in the food and by the carbon monoxide in tobacco smoke and motorcar exhaust can no longer be dismissed as trivial.

Sodium nitrite is highly toxic in relatively small amounts. About four grams of the compound constitutes a lethal dose for adults. Although Lehman regards 200 parts per million of sodium nitrite as a safe residue, he notes that "only a small margin of safety exists between the amount of nitrite that is safe and that which may be dangerous. The margin of safety is even more reduced when the smaller blood volume and the corresponding smaller quantity of hemoglobin in children is taken into account. This has been emphasized in the recent cases of nitrite poisoning in children who consumed weiners and bologna containing nitrite greatly in excess of the 200 ppm permitted by Federal regulations. The application of nitrite to other foods is not to be encouraged."

Emulsifiers have been used in bread not only as softening agents, which can give stale bread the texture of newly baked bread, but as substitutes for nourishing ingredients. "The record of the bread-standards hearings contain evidence of distribution among bakers of advertising material advocating the replacement of fats, oils, eggs and milk by emulsifiers," George L. Prichard, of the Department of Agriculture, told the Delaney Committee. "The use of such products as components of food may work to the disadvantage of our farm economy by displacing farm products normally used. The record indicates that natural food constituents, such as fats and oils, probably will be reduced in many commercial bakery products if bakers are allowed to employ these emulsifiers."

This substitution affects more than 117 the 117 farm 117 economy, however; it also works to the disadvantage of the consumer. To illustrate this, the Delaney Committee report compared the ingredients in two cake batters prepared by the same company eleven years apart; during the interval emulsifiers were added to the product. The first batter, made in 1939, did not contain synthetic emulsifiers; the second, prepared in 1949, did. "On a percentage basis, the cake batter in 1939 contained 13 per cent eggs and 8.6 per cent shortening.

In 1949, the cake batter contained 6.3 per cent eggs and 4.8 per cent shortening, with somewhat less than 0.3 per cent of synthetic emulsifier." The report noted that a "synthetic yellow dye could be added to provide the colour formerly obtainable through the use of eggs. The utilization of synthetic yellow dye in commercial cake was practiced before the war. There are indications that the use of artificial coloring matter is increased when quantities of whole eggs or egg yolks are reduced in commercial cake formulas."

To heighten the insult, among the most commonly used emulsifiers in 1949 were the polyoxyethylene monostearates. The yellow dye referred to in the Delaney Committee report was probably FDC Yellow No. 3(Yellow AB), which, until fairly recently, was added to many yellow cake mixes. As we shall see in the following chapter, the dye often contained impurities of a potent carcinogen and its use in foods was forbidden in 1959. For a number of years, however, both the emulsifier and the dye undoubtedly appeared in many brands of cake and reached large numbers of unsuspecting consumers.

Problems of this kind are not likely to disappear unless there is a basic change in the viewpoint of the F.D.A. "Inherited from the Wiley era is a too common misconception that all 'chemicals' are harmful and the related idea that any amount of a 'poison' is harmful," the F.D.A. observes in a brochure on food additives. "The fact is, of course, that chemical additives, or food *additives* as they are now being called, have brought about great improvements in the American food supply. Additives like potassium iodide in salt and vitamins in enriched food products are making an important contribution to the health of our people and yet it is a fact that both iodine and some of the vitamins would be harmful if consumed in excessive amounts. Many similar examples could be given to refute these common misconceptions."

This argument is grossly misleading. Iodine and vitamins are indispensable to human life, but coal-tar dyes and benzoic acid, for example, are not. If we adhered to a well-balanced diet of properly prepared natural foods, iodine and vitamins would never enter the body in toxic amounts. Coal-tar dyes, on the other hand, are suspected of being harmful to man in nearly any amount if consumed repeatedly and the kindest statement that can be made for the presence of benzoic acid in food is that the compound is "safe under the conditions of its intended use."

The formula "safe under the conditions of its intended use" exposes the consumer to serious risks when it opens the door to the use of food additives whose biochemical activity is not understood. Many unexpected problems may arise when such additives appear in food. A particular additive may seem to be relatively harmless to the organs of the body, but it may be carcinogenic on the cellular level of life.

Another additive may produce insignificant or controllable effects when studied in isolation; brought into combination with various chemicals in food, however, it may give rise to toxic compounds. The body, in turn, may make a toxic additive more poisonous in the course of changing it during metabolism. "At one time it was generally believed that whenever a toxic substance was absorbed the body was capable of calling upon special mechanisms for detoxifying the toxicant," Lehman observes.

"It was believed also that the metabolic pathway that the toxicant followed always proceeded in the direction of the formation of a less toxic compound. Later work on the metabolism of drugs and toxic substances showed that special mechanisms did not exist, but that the toxicant was subject to the same metabolic influences as those which normally operate in the body.

The assumption that the metabolic product was less

poisonous than the parent substance from which it was derived is also unwarranted simply because in many instances the toxicity of either the original substance and its conversion product is unknown. In other instances the metabolic product is even more poisonous than its parent. The conversion product, heptachlorepoxide, which is two or three times more poisonous than the parent substance, heptachlor [a widely used insecticide], may be cited as an example of this."

Finally, food additives may cause allergic reactions that are likely to go unnoticed for many years. The reactions need not be severe to be harmful. Otto Saphir and his colleagues at the Michael Reese Hospital in Chicago have recently suggested that the development of arteriosclerosis may be promoted by the sensitivity of the body to allergenic compounds, notably certain antibiotics. By using sulfa drugs to produce allergies in forty-two rabbits, the researchers were able in eight months to cause degenerative changes in the arteries of thirty-one of the animals.

These changes closely resembled arteriosclerosis in man. According to a press account of Saphir's report, the rabbits' reactions "were not apparent on the surface." It would be very imprudent to assume that such effects are produced only by chemicals that cause severe or noticeable allergies. If the data of Saphir and his colleagues are applicable to man, additives with even minor allergenic properties cannot be dismissed as harmless.

Today more than 3,000 chemicals are used in the production and distribution of commercially prepared food. At least 1,288 are purposely added as preservatives, buffers, emulsifiers, neutralizing agents, sequestrants, stabilizers, anticaking agents, flavoring agents and coloring agents, while from 25 to 30 consist of nutritional supplements, such as potassium iodide and vitamins.

The remaining chemicals are "indirect additives"—substances, such as detergents and germicides, that get into food by way of packaging materials and processing equipment.

Many chemical additives are natural ingredients, but a

large number are not. The artificial substances that appear in food range from simple inorganic chemicals to exotic compounds whose biochemical activity is still largely unknown.

At a time when almost every processed food contains chemical additives or residues, the "misconceptions" of the "Wiley era" must seem like bold anticipations. Sixty years ago the number of chemicals added to food was small. Public concern was aroused primarily by the gross adulteration of food and by the unsanitary practices followed in mills and slaughterhouses.

These problems are still with us; as recently as 1958, over 5,000 tons of food was seized by the F.D.A. because of filth and decomposition.

But the problem of chemicals in food has reached proportions that would have appalled Wiley and the incidence of diseases, such as cancer, that can be produced by chemicals in man's environment has increased to an alarming extent. We sense that our food technology has taken a wrong turn. Having achieved the abundance of nutriment it once promised, it threatens to become another factor imperiling man's health and well-being.

Chapter 3

Effects of Radiation

Radioactive elements are among the most lethal toxicants in man's environment. As little as four micrograms (one seven-millionth of an ounce) of radium is capable of producing cancer in man years after lodging in his bones. Not all radioactive elements are dangerous in the same way as radium. Some are distributed differently in the body and are not retained for long periods; others, although retained for years, emit different kinds of radiation. Strontium-90, for example, is rated about one tenth as lethal as radium, although it, too, lodges primarily in the bones. Attempts to compare the potency of the two elements in quantitative terms, however, can be somewhat misleading.

Radium causes biological damage through the emission of alpha particles, whereas strontium-90 is hazardous to living things because it is an energetic emitter of beta particles.It is Hardly necessary to emphasize that since the explosion of a nuclear weapon at Alamogordo, New Mexico, in July 1945, ionizing radiation has become the most serious threat to man's survival. If nuclear weapons are further developed or increased in number, it is quite conceivable that a future war will virtually extinguish human life. Grave as this danger may be, ionizing radiation is associated with a number of subtle hazards that warrant equal public concern.

"Today sources of ionizing radiation are rapidly becoming more and more widely used," observes a team of radiation specialists for the U. S. Public Health Service. "Each atomic reactor produces the radioactive equivalent of hundreds of pounds of radium. Radioactive substances are used in increasing numbers of hospitals, industries and research establishments. X-ray machines are encountered in tuberculosis surveys, in shoe stores, in physicians' offices and in foundries. Potentially, ionizing radiation has come to affect the environment of every member of the community."

"Leukemia has been observed to occur in radium patients, though not as a prominent phenomenon," observes W. F. Neuman, of the University of Rochester. "Leukemia is, however, a common finding in people who have suffered irradiation of the bone marrow or of the whole body. Since the beta rays from strontium-90 are more penetrating than the alpha rays from radium, a greater proportion of the rays from strontium-90 will penetrate into the marrow." It is quite possible that, for individuals prone to leukemia, strontium-90 poses hazards in addition to those that would be expected on the basis of criteria derived from experiences with radium.

To gain a clear understanding of the various kinds of radiation and their biological effects, it will be useful to review briefly some elementary concepts of atomic physics. Radioactivity is produced by changes that occur within the nucleus of an atom. These changes either directly or indirectly involve two nuclear particles, the neutron and the proton. The number of neutrons (electrically neutral particles) and protons (positively charged particles) in an atomic nucleus determines the species to which the atom belongs.

Carbon-12 and carbon-13, for example, have six and seven neutrons, respectively and are called isotopes of the element carbon. Every carbon atom, however, has six protons in its nucleus and an equal number of external electrons (negatively charged particles), which can be thought of as "revolving" around the nucleus like planets in a miniature solar system. Similarly, all hydrogen atoms have one proton and one planetary electron, whereas hydrogen-2 and hydrogen-3 have one and two neutrons, respectively.

When carbon and hydrogen combine to form hydrocarbons, the nuclear particles (protons and neutrons) of the two elements are not affected by the combination. The chemical reaction of carbon and hydrogen is determined by the arrangement of their planetary electrons. Each element can be separated from the other without any loss of its original identity and without producing ionizing radiation. On the other hand, if carbon were to gain or lose any protons, it would become an entirely different element.

Gains or losses in the number of protons may occur when a neutron is converted into a proton, when a proton is converted into a neutron and, of course, when a proton together with a neutron is emitted from the nucleus of an atom. These changes and the various energy adjustments that occur within the nucleus produce ionizing radiation. The more familiar elements that make up the earth do not undergo nuclear changes spontaneously. Their nuclei are extremely stable.

A few elements, however, such as uranium, radium and polonium, have unstable nuclei. They emit particles and eventually break down into more stable elements. Radium, which is derived from the decay of uranium, disintegrates into radon by emitting two protons and two neutrons. Decay continues through a series of radioactive daughter products until the atom becomes lead. The two protons and two neutrons emitted by the radium atom remain together and constitute an alpha particle.

Streams of these neutronproton groups, or alpha "rays," are emitted from radium nuclei at velocities of about 44,000 miles per second. In addition, radium emits gamma rays and several daughter products of the element emit beta rays. Beta rays are streams of electrons and positrons (the positive counterparts of electrons); they travel at varying speeds, sometimes approaching the speed of light (186,000 miles per second). Gamma rays are physically the same as X rays, the difference in name arising from the difference in the sources of emission.

It takes 1,600 years for half a given quantity of radium to disintegrate into radon. Another 1,600 years must pass before the remaining quantity of radium is halved again. As this process goes on indefinitely, always leaving a remainder to be halved each 1,600 years, some radium will remain for an immensely long period. By the same token, by adding radioactive substances to his environment, man increases the amount that will remain indefinitely, whether the halflife of an element is 5,900 years (carbon-14) or 28 years (strontium-90).

Although man can create radioactive elements such as strontium-90, there is nothing he can do to reduce their radioactivity. Radiation continues undisturbed and is diminished only with the passage of time. What effects do alpha, beta and gamma rays produce in living tissues? Although the long-range effects of low-level radiation on animals and man are very much in dispute, there seems to be little doubt about the harmful changes radiation produces in cells. Radiation particles have been compared to bullets tearing into an organism, but the damage they cause depends upon the number of particles that reach the tissues.

The particles interact with many of the atoms that make up our cells, either causing electrons to leave their orbits (ionization) or exciting combinations of atoms (molecules) to a point where they break apart. The resulting damage may be minor and easily repaired; on the other hand, it may be serious and permanent. Irradiated germ cells may one day give rise to harmful mutations and irradiated body cells may take the first step toward the development of cancer. An individual exposed to large dosages of radiation may exhibit symptoms of "radiation sickness"—nausea, vomiting, loss of hair, progressive weakness and hemorrhages.

Severe irradiation of the entire body damages the blood-forming organs in man and higher animals, often claiming life within several weeks after exposure. If the whole body is exposed to very high levels of radiation, there will be widespread destruction of the intestinal tract and nerve cells. In such cases, death is certain to occur within a few days after exposure. Quantities of radiation are usually expressed in terms of rads and roentgens.

A rad denotes the quantity of radiation an individual absorbs; a roentgen denotes the amount of radiation (ordinarily X radiation) to which he is exposed. For the purposes of this discussion, no significant distinction need be made between the two units. A single rad or roentgen can be thought of as giving rise to about 1,000 ionizations in a single cell.

An individual normally receives a total dosage of 3 to 5 roentgens from natural background radiation during the first thirty years of his life. Background radiation comes from cosmic rays, radioactivity in rocks and radioactive elements in food and water. It constitutes the radiation to which man has evidently adapted himself over long periods of biological evolution. Although background radiation will vary in amount from one locale to another, the absolute quantities involved are ordinarily very small. A man living at sea level usually absorbs about 30 millirads (30 thousandths of a rad) of cosmic radiation a year. A man living at 5,000 feet above sea level is likely to absorb about 70 millirads annually.

An individual who moves from sea level to a height of 5,000 feet will thus raise his absorption of cosmic radiation by little more than a rad during a thirtyyear period. A man who remains behind at sea level may well acquire more than that amount in certain parts of his body, by having a tooth X-rayed, for example. In fact, a series of dental X rays will supply his jaw and neck with a larger amount of radiation than they are likely to acquire from natural background sources in the greater part of a normal life span. When X rays are used for diagnostic or therapeutic purposes, the good they do outweighs the harm.

Diagnostic X rays are usually directed at limited areas of the body, a form of exposure that is far less hazardous than irradiation of the whole body. However, despite recent claims to the contrary, there is no conclusive evidence that small doses of radiation are harmless. Although some research workers have reported that very low levels of radiation (5 roentgens or less per week) increase longevity in animals, others working with similar doses of radiation have found that the life span is shortened.

In the cases in which increases in longevity were reported, the last animals to die were generally non-irradiated control animals. If radiation had any "beneficial" effect, it was that it seemed to reduce the death rate of irradiated animals in early age groups, but the maximum life span of the animals was shortened. There seems to be very little ambiguity, however, about the genetic effects of radiation. All the facts at hand indicate that any increase in the amount of radiation received by a large population produces a rise in the occurrence of harmful mutations.

Nor do we have to rely exclusively on experiments with animals to acquire statistically significant evidence to support this conclusion. A study made by John T. Gentry and his colleagues in the New York State Department of Health suggests that areas with relatively high concentrations of radioactive rock are correlated with a high incidence of congenitally malformed children many, perhaps most, of whose defects are believed to be caused by defective genes.

By working primarily with statistics compiled from a million and a quarter birth certificates in New York State, Gentry found that the over-all incidence of congenital malformation amounted to 12.9 per 1,000 live births in areas of the state which seemed to have the least amount of geologic radioactivity. The incidence of congenital malformation for areas which seemed to have the highest amount of geologic radioactivity was 17.5 a difference of nearly 40 per cent.

New York State, it should be noted, is not particularly radioactive; field measurements of background radiation were found to lie mainly in the range of 2.1 to 3.2 roentgens per thirty-year period. Additional evidence that the rate of malformation is influenced by the radioactivity of the geologic environment appears in a preliminary nation-wide study by Jack Kratchman and Douglas Grahn, of the U. S. Atomic Energy Commission.

By grouping county and state mortality data according to the geologic provinces of the United States, Kratchman and Grahn found that "mortality incidence from malformation may be higher in those geologic provinces that contain major uranium ore deposits, uraniferous waters or helium concentrations."

These studies are a reminder that man is probably far more sensitive to radiation than has generally been supposed.

Although the living matter that surrounds the nucleus of a cell can often absorb a great deal of radiation without being significantly harmed, the nucleus itself is highly vulnerable. Theodore T. Puck and his coworkers at the University of Colorado Medical Centre have recently shown that visible damage can be inflicted on chromosomes (the strands in the nucleus of the cell that carry the genes) with as little as 20 to 25 roentgens.

The genetic apparatus of an ordinary body cell "thus shows extraordinary sensitivity to radiation," Puck observes. "Calculation indicates that the ionization of not more than a few hundred atoms within or around the space occupied by the chromosome may be sufficient to produce a visible break. It seems probable that an even greater number of lesions may occur and remain invisible, either because they are submicroscopic or because they become resealed before they are expanded." If radiation is increased to 50 roentgens, many cells soon lose the ability to reproduce, although they do not necessarily die. The nuclear material of the cell may be damaged beyond recovery, but the cytoplasm surrounding the nucleus survives and performs the essential functions of life.

Instead of reproducing, the cells continue to absorb nutriment and grow until they reach enormous size. They become so-called "giant cells," seven to ten times larger than a cell that has not been irradiated. Irradiation of the whole body with 400 to 500 roentgens is fatal to half the human beings so exposed, apparently because over 99 per cent of the body's reproducing cells lose the ability to multiply normally. Individuals who receive lethal dosages of radiation may live for weeks after exposure, but the body's inability to form a sufficient number of red and white blood cells eventually leads to death from anemia or infection. Children are probably more sensitive to radiation than adults and the fetus appears to be more sensitive than the child.

Radiation not only has a selective effect upon different parts of a cell; it does the greatest damage to those cells that have a high rate of reproduction. It seems to be most harmful during those stages of life at which metabolism is at its peak, notably the years of growth. Sensitivity varies markedly from individual to individual and from one organ of the body to another. A large whole-body dose of radiation from which some individuals recover with a self-limiting case of "radiation sickness" may lead in others to permanent damage and death. If concentrated on the hand, the same dose may produce nothing more than a radiation burn.

Directed at the abdominal area, however, it will often cause serious damage and marked physiological reactions. Finally, there are different kinds of radiation and various emitters of radiation. The effects produced by all forms of radiation are ordinarily reducible to the amount of ionization they produce. But each form of radiation has distinctive features of its own. The alpha particle is the "heavy artillery shell" of ionizing radiation. As it is made up of two protons joined with two neutrons (actually a nucleus of the element helium), an alpha particle has about 7,300 times the mass of the lighter, faster beta particle.

The penetrating power of alpha rays is very limited; they can be stopped by a thin sheet of paper. But although alpha particles do not penetrate very far into living tissue, they score a large number of "direct hits" on atoms. An irradiated cell is damaged more severely by alpha particles than by other forms of radiation, although fewer cells are irradiated by an alpha emitter. If the alpha particle had the penetrating power possessed by beta and gamma rays, it would be extremely destructive to living things.

Beta particles travel about 100 times farther than alpha particles. Whereas alpha particles seldom pass beyond the outer, dead layer of the skin, the free, fastmoving electrons and positrons that constitute beta radiation penetrate for about a quarter of an inch into living matter. Gamma rays and X rays will pass readily through a large organism; they reach the innermost recesses of the body and injure highly sensitive tissues, but they produce only about one twentieth of the damage inflicted on cells by alpha particles.

As gamma rays are physically the same as X rays, we might say that most of the effects produced in man by gamma radiation are due to the widespread use of X-ray equipment. Alpha and primary beta radiations are carried to man's vital organs by radioactive elements. X-ray equipment can be turned on and off; thus, exposure to X radiation need not be continuous or uncontrolled. But radioactive isotopes cannot be "turned off." Once they combine with the skeleton and the soft tissues of the body, they are no longer within human control.

Like the nutrients that sustain life and growth, many radioactive elements become part of the organism. The body tends to use radium and strontium-90, for example, the same way that it employs calcium, although it shows a preference for the latter. Radium and strontium-90 migrate to the mineral part of the skeleton, where the rate of metabolic activity is relatively low and they bombard bone matter and highly sensitive bone marrow with alpha and beta particles for a long period of time. The body does not discriminate between radioactive and stable carbon isotopes.

Carbon-14 appears in every kind of tissue, including germ cells and irradiates the area in which it lodges with beta particles. In the field of radiobiology, it is very imprudent to make ironclad generalizations that are likely to affect the well-being of large populations. The field is beset with so many uncertainties that scientific opinion can be expected to differ on nearly every issue but the "need for further research." The public, in turn, would probably acknowledge this need with tranquillity if the problems of radiation were confined to the laboratory. But they are not.

X-ray machines are now part of the equipment of every physician and dentist; the devices have become indispensable in medical practice. Radioactive isotopes are used widely in laboratories, hospitals and industry. Nuclear weapons tests have discharged radioactive debris into the soil, water and atmosphere and nuclear reactors are annually producing huge quantities of highly radioactive wastes.

The experimental animals needed for further research currently include all of humanity and the modern radiation laboratory encompasses the entire earth. But many serious questions have been left unanswered. Are we using radiation with the caution and respect it deserves? To what extent have we contaminated the environment? In what direction are we going? Unless these questions are answered frankly and existing abuses corrected, the data acquired by further research may well prove to be of academic importance.

THE PROBLEMS OF X RADIATION

X-rays were the first form of ionizing radiation to be deliberately produced by man and in the United States and western Europe they now constitute the radiation to which he is most heavily exposed. Many people of all ages and all walks of life are given routine X rays during periodic medical and dental examinations. Moreover, exposure to X rays has been increasing steadily over the past three decades. According to data gathered by the National Advisory Committee on Radiation, the annual X-ray dosage received by the average American increased 900 per cent between 1925 and 1955.

The radiation he receives from X rays is 135 per cent greater than the radiation he acquires from natural sources. "The continued upward trend exhibited by X-ray data," observes the committee, "suggests the likelihood that the current exposure of the population from X-ray apparatus may increase still further unless appropriate radiation control measures are systematically applied."

Systematic control of X radiation is long overdue. Man-induced radiation has been in use since 1896, the year Roentgen's discovery of X rays attracted world-wide attention and the effective application of these rays to medical problems antedates World War I. Surprisingly, a good working knowledge of the hazards of X radiation and valuable recommendations for the proper use of X-ray equipment were available shortly after Roentgen's work became well known. Elihu Thomson, after experimenting with the new rays on his little finger, reported a great deal of the data needed to enable radiologists to use X-ray equipment with a reasonable amount of safety. His suggestions were largely ignored. For many years X-ray tubes were employed with little or no shielding, while physicians, in blissful disregard of Thomson's recommendations, often exposed themselves as well as their patients to heavy doses of radiation. By the early 1920's, scores of radiologists had needlessly sacrificed their lives and, in all probability, inflicted a great deal of damage on many of their unsuspecting patients.

To make matters worse, X-ray equipment was rapidly debased into a cosmetic agent and, finally, into a salespromotion device. It was found that X rays could cause a loss of hair (epilation), an effect that suggested lucrative possibilities. By the 1920's many physicians, beauticians and self-appointed "epilation specialists" had begun to treat women with radiation for the removal of "superfluous hair." One New York physician, Dr. Albert C. Geyser, developed a "harmless" method of hair removal that involved cumulative dosages of at least 500 roentgens over a twelve-week period of radiation treatment.

The method, named the "Tricho System," was very successful and beauticians trained by Geyser's "Tricho Institute" began operating in many parts of the United States and Canada. It soon became evident, however, that women treated according to the "Tricho System" lost substantially more than unwanted hair. Many individuals acquired radiodermatitis (skin inflammation), severe radiation burns, festering skin ulcers and, in time, cancer. The "Tricho" story is one of the more tragic episodes in the history of radiation. It is believed that the victims of Geyser's system numbered in the thousands; the exact number of those who suffered latent injury and premature death will never be known.

Although radiation is no longer employed in the American beauty parlor, the use of X-ray equipment to fit shoes still lingers in a number of communities. The equipment is used mainly on the feet of children. As of 1960, the use of the shoe-fitting fluoroscope had been banned in twenty-nine states. Some of the other states regulate the use of the machine, but in a few states there are no restrictions at all.

A number of local surveys cited by Schubert and Lapp have shown that the machines are often defective, giving high doses of radiation to both the child and the salesman. The Michigan Department of Health, for example, found shoe-fitting machines that emitted as many as 65 roentgens (r) a minute. A survey in Boston showed that irradiation of the foot ranged from 0.5 to 5.8 r a second. "For a 22-second exposure, which is commonly used, the feet receive from 10 to 116 r!" Schubert and Lapp write.

"Remember, too, that one child may have his feet examined many times while trying on different shoes. Similar dosage measurements have been reported by the United States Public Health Service, which states that the average dosages to the children's feet are between 7 r and 14 r per exposure." The amount of scattered radiation that reaches the child's pelvic region and gonads may run as high as 0.2 roentgens for a twentysecond exposure.

Society has found it very difficult to accept the fact that a valuable device can become extremely dangerous if it is used improperly. Schubert and Lapp observe that a major reason why Geyser was not stopped until his "system" had injured a large number of women is that the medical community generally accepted radiation as a form of treatment for minor skin disorders. To this day, many physicians are likely to underestimate the amount of latent damage caused by repeated exposure to X radiation. In some cases, physicians are inclined to use X-ray equipment as freely as they use the stethoscope and often it is the cost of an X-ray picture rather than the risk of irradiation that keeps the two devices from occupying places of equal importance in routine medical examinations.

The physician often feels that there is no danger involved in taking diagnostic X rays; the word "diagnostic" seems to impart benign qualities to radiation. But the need for an X ray does not in any way diminish the effect of exposure; it merely provides a scale on which the hazards of X radiation should be weighed against the hazards of an incomplete and faulty diagnosis. This requires careful evaluation.

A middle-aged man who refuses to be exposed to the small amount of radiation involved in an annual chest X ray carries his fear of radiation to extremes. The risk involved in having an X ray taken is much smaller than the risk involved in permitting a serious lung disease to remain undiagnosed.

Schubert and Lapp estimate that sixty chest X rays taken on a 14-by-17-inch plate without accompanying fluoroscopies will give an individual a cumulative dose of 3 roentgens. With newly developed fast films, exposure is likely to be reduced appreciably. Similarly, a man with marked and prolonged gastrointestinal distress who fails to respond to treatment for benign stomach disorders would be foolhardy if he refused to have X rays taken of his abdominal region. The danger of serious disease would outweigh by far the amount of harm caused by a single series of X rays.

But individuals without medical complaints who are exposed periodically to a large amount of diagnostic radiation might do well to entertain some second thoughts. For example, many large American corporations have established the practice of sending their executives to clinics for annual medical examinations. These examinations take several days to perform and normally include a programme of X rays and fluoroscopies of important organs of the body. Schubert and Lapp found that an executive can expect to receive 35 roentgens with each check-up; frequently the dosages reach 50 roentgens.

"We grant that the executive is not exposed to total body radiation, which is more serious than localized radiation," they write. "Nevertheless, the most vital organs are exposed and at the very least one may expect a shortening of life span of the exposed individual, especially if the executive receives the x-ray bombardment year after year. We do not deny the valuable data which doctors may gain in the course of x-ray examination, which may lengthen the life span, but we do regard the annual irradiation of 50 r per executive with some degree of apprehension for the individual's future welfare."

A number of radiobiologists believe that certain X rays should not be undertaken unless they are absolutely necessary for the well-being of the patient. A case in point is the pelvic X ray (pelvimetry) of pregnant women. "X-ray pelvimetries have been a relatively common practice for many years—and still is as far as is known," Schubert told a congressional committee holding hearings on radiation in 1959. An estimated 440,000 American women and 100,000 English women were given pelvic X rays annually during the late 1950's.

The practice had always aroused a certain amount of concern because of the harmful genetic effects it could produce in later generations, but recent studies show that the hazards may be more immediate. In an extensive survey of deaths from cancer among English and Welsh children, Alice Stewart and her co-workers at Oxford University found a statistical relationship between fetal irradiation and the incidence of cancer among children. The survey, perhaps the most extensive of its kind, covered most of the children who had died of leukemia and other cancers between 1953 and 1955.

The data gathered by the investigators show that, among children under ten, the chances of dying from cancer are twice as great for those who were irradiated during the fetal stage. Moderate enlargement of the thymus gland, a condition for which a large number of children have received radiotherapy, is now regarded as harmless by many medical authorities. Although thymic enlargement is no longer treated with radiation in England, the practice still persists in the United States. C. Lenore Simpson and her co-workers have carefully surveyed about 2,000 such cases in New York State, mainly in Rochester and Buffalo.

By comparing the incidence of cancer in treated cases with that in children in the general population, they found nearly a sevenfold increase in cancer among children who had been exposed to high dosages of radiation for thymic enlargement. The number of leukemia cases found by Simpson was nearly ten times higher than the number that would normally have been expected. It remains to be seen whether most radiobiologists will decide that pelvimetries and thymic irradiation produce cancer in children.

The evidence is not altogether clear-cut. Several later surveys of a more limited nature have reported findings that conflict with the results obtained by Stewart and Simpson, whereas other surveys support them. The results seem to vary with the methods used and the communities studied.

The Stewart and Simpson surveys, however, have unearthed statistical probabilities that are far too high to be dismissed lightly. Although these probabilities can be expected to vary with the techniques of the radiologists in different communities and medical institutions, there seems to be little reason to doubt the validity of the findings of Stewart and Simpson. Physicians have observed that many children with thyroid cancer received radiation around the head and neck during infancy and early childhood.

George Crile, of the Cleveland Clinic Foundation, for example, reports that in 18 cases treated by his clinic, the parents of 14 were asked whether their children had been exposed to X rays in the general area of the thyroid gland. Only 3 had not received radiation.

The remaining 11 had histories of radiotherapy for enlargement of the thymus gland and lymph nodes, for eczema and for other disorders. "In the 15 years between 1924 and 1939," Crile notes, "at a time when many more operations on the thyroid were being done at the Cleveland Clinic than are being done today, no children with cancers of the thyroid were seen. Only three were seen between 1939 and 1950 and between 1950 and 1958 there were 15 cases. The question immediately arises as to whether the increasing incidence of cancer of the thyroid in children is the result of an increase in the use of radiation therapy in infancy and early childhood."

Crile notes that factors other than radiation may account for the rise in thyroid cancer among children but adds that "the increasing incidence of cancer of the thyroid in recent years suggests that an environmental factor such as radiation is involved." Many individuals are still exposed to large dosages of X radiation for what are essentially cosmetic purposes. Localized eczemas, warts and childhood birthmarks that are likely to disappear spontaneously after a few years have in many cases been heavily irradiated. This practice is declining owing to a greater awareness of radiation hazards, but it is still found too often to be overlooked.

Radiation, to be sure, is still the only effective method for treating certain irritating or disfiguring skin disorders and the patient may be the first to demand radiotherapy in full awareness of the risk involved. Radiotherapy also reduces the severe pain that accompanies ankylosing spondylitis, an arthritic disease of the spine that afflicts thousands of men and often leads to a useless or impaired life. In such cases, where irritation or pain may be severe, it is unlikely that the hazards of radiation will deter a patient from accepting radiotherapy. He will probably take the risk—especially if the risk is relatively small.

The element of risk cannot be avoided, but the amount of risk can be reduced appreciably. X-ray equipment can be used with moderation and with good sense. Doctors should be made thoroughly aware of the hazards involved and they should be taught the most advanced methods of reducing exposure. The physician should limit the use of his equipment to those aspects of radiology with which he is thoroughly familiar. A general practitioner is not a radiologist. Where complex techniques are required to diagnose a disease or treat a patient, the patient should have the benefit of the special training and experience that come from long service in the field of radiology.

There is also an area which both layman and physician can enter with equal authority—that of radiation safety. Careful shielding from scattered X rays should always be provided for sensitive areas of the body, such as the gonads, neck and abdominal region, particularly when children and young people are being irradiated. A patient has a right to insist upon protection whenever it is feasible and in this he has the emphatic support of the most knowledgeable authorities in the field of radiology.

The use of X-ray equipment should be carefully regulated. These devices do not belong in the hands of quacks and shoe salesmen. Technicians should be licensed personnel who have given substantial evidence of their qualifications to operate X-ray equipment. Their work requires careful training that cannot be picked up through irregular, offhand instruction.

There should be compulsory periodic inspections of X-ray equipment by competent agencies. Concerted efforts must be made to bring the latest advances in radiology into physicians' offices and hospitals.

Research has produced faster films, electronic devices to increase the brightness of fluoroscopic screens (with concomitant reductions in X-ray dosage), image intensifiers and improved filters that eliminate diagnostically useless long-wave radiation. Some of these improvements are too costly for the ordinary physician; hence the need to make use of the services of a well equipped radiologist or hospital when a programme of irradiation is required. It is the responsibility of the community to see that outdated X-ray equipment is scrapped and that no one is exposed to defective and potentially harmful machines.

Unfortunately, the available evidence suggests that, in most cases, both the machines and the physicians' techniques are unsatisfactory. A recent two-year survey of diagnostic X-ray equipment in New York City, for example, showed that 92 per cent of 3,623 machines inspected by the Board of Health either were not being used properly or were defective. The survey disclosed that X-ray beams were very broad, needlessly irradiating parts of the body that were not under study.

The majority of physicians who were not radiologists were unfamiliar with the safety recommendations of the National Committee on Radiation Protection. The inspectors did not find a single physician who had voluntarily followed earlier recommendations to switch from outmoded to new equipment. Although the survey covered only 35 per cent of the machines employed in the city, it included the most frequently used X-ray machines, notably those in hospitals and in the offices of radiologists.

This survey, it should be emphasized, took place in 1959 and 1960, not a half century ago, when research on radiation was still in its infancy. The survey followed a period of widespread public discussion on the hazards of radiation and nuclear fallout. It is now clear that the situation is much worse than had generally been supposed. Exposure to radiation is occurring on a scale that has no precedent in man's natural history. Millions of people in all stages of life, from the fetal to the senile, are being irradiated every year.

Clearly the problem of radiation control has reached serious proportions. From the various reports "on the influence of ionizing radiation on biological systems," observes the National Advisory Committee on Radiation, " it is evident that serious health problems may be created by undue exposure and that every practical means should be adopted to limit such exposure both to the individual and to the population at large."

X-RAY EQUIPMENT

X-ray Equipment, we noted earlier, can be turned on and off, but the radioactive wastes that enter man's environment through nuclear weapons tests and the activity of nuclear reactors are essentially beyond human control. They contaminate air, water and food and they irradiate everyone, irrespective of age or health. Radioactive contaminants also create problems not encountered with conventional pollutants.

Ordinary contaminants usually lose their toxic properties by undergoing chemical change, but there is no loss of radioactivity involved in the chemical reactions of radioisotopes. When radiocarbon combines with oxygen to form carbon dioxide, the carbon in the compound continues to emit beta particles. The same is true for chemical compounds formed by strontium-90. Radioactivity persists in all radioisotopes until unstable atoms decay into stable ones.

Until recently, the layman was given a highly misleading picture of the hazards created by nuclear weapons tests. This picture was largely created by the Atomic Energy Commission, the official agency that had been made chiefly responsible for furnishing the public with information in the field of nuclear energy. For many years the A.E.C. consistently minimized the danger posed by radioactive fallout produced by nuclear weapons tests. For example, it completely ignored the extent to which food had been contaminated with strontium-90 until the problem was raised by scientists who were critical of the agency's public information policies.

"In the 13th Semiannual Report of the AEC, published in 1953," notes Barry Commoner, of Washington University, "the AEC stated that the only possible hazard to humans from strontium-90 would arise from 'the ingestion of bone splinters which might be intermingled with muscle tissue during butchering and cutting of the meat.' No mention of milk was made"—or, for that matter, of vegetables and cereals.

Spokesmen for the A.E.C. predicted that fallout would be uniformly distributed over the earth, so that no areas need fear concentrations of debris from nuclear weapons tests. The public was assured that the greater part of the debris sent into the stratosphere would remain aloft for a period of five to ten years. As fallout occurs very slowly, it was said, the radioactivity of short-lived radio-isotopes would be almost entirely dissipated in the stratosphere.

Actually, the radioactive debris that soars into the stratosphere stays there, on an average, less than five years. According to General Herbert B. Loper, of the Department of Defence, half the stratospheric debris produced by a nuclear explosion returns to the earth within two years. Fallout occurs three and one half times faster than Willard F. Libby, former commissioner of the A.E.C, had estimated.

A model of stratospheric air circulation developed by A. W. Brewer and G. M. B. Dobson indicates that the heaviest fallout in the Northern Hemisphere occurs in the temperate zone, reaching a peak between 40 and 50 degrees north latitude—or roughly between Madrid and London in Europe and between New York City and Winnipeg in North America. Measurements made during the 1958-61 nuclear weapons test moratorium indicate that the hazard from fallout in these latitudes is substantially greater than the world-wide average.

The rapidity with which radioactive debris descends to the earth places the danger presented by short-lived, supposedly harmless radioactive elements in a new perspective. Cesium-144 and strontium-89 have half-lives of only 290 and 56 days, respectively, but nuclear explosions produce these radioactive elements in such relatively large quantities that, if fallout is rapid, they become a serious hazard to public health. Cesium-144, like long-lived cesium-137 (another component of fallout), is an emitter of beta rays. When taken into the body, both cesium isotopes are handled metabolically like potassium; they migrate to all the soft tissues, including the reproductive organs.

Strontium-89 possesses the characteristics of strontium-90; it, too, emits beta rays and tends to lodge in bone matter. Although a short-lived bone seeker like strontium-89 might seem to be relatively harmless, it should not be underestimated as a hazard to public health. "Since strontium-89 is produced more abundantly in fission than strontium-90," the Special Subcommittee on Radiation of the Joint Committee on Atomic Energy reported in 1959, "it is possible that comparable doses to the body from the two materials could occur." The subcommittee added that "it would require 100 times more initial activity of strontium-89, whose half life is 56 days, to deliver the same dose to tissue that would be created by I unit of strontium-90. It is considered significant that transient levels of strontium-89 with approximately this ratio to strontium-90 have been observed in milk."

For a few weeks after a nuclear explosion, the windborne debris in the lower part of the atmosphere may contain appreciable amounts of iodine-131. Iodine-131 has a half-life of eight days. At the 1959 hearings of the Special Subcommittee on Radiation, E. B. Lewis, of the California Institute of Technology, observed that "the radioiodines in fallout are a special hazard to infants and children. This hazard arises for a variety of reasons. Radioiodine is a significant fraction of the fresh fission products released by nuclear weapons explosions.

Grazing cattle ingest and inhale the radioiodines in fallout and then concentratate it in their milk. Infants and children are expected to ingest more of the isotope than will adults since fresh cow's milk is the principal source of fallout radioiodine in the human diet and young people obviously drink more fresh milk than do adults. As has long been known, iodine isotopes, natural and radioactive, concentrate in the thyroid gland. Moreover, for the same amount of radioiodine orally ingested, the infant thyroid receives some 15 to 20 times the dose that the adult thyroid receives.

(Briefly, this is because more radioiodine is taken up by the infant than by the adult thyroid; as a result many more of the short-ranged iodine-131 beta rays will be generated in a gram of infant than in a gram of adult thyroid tissue.) Finally, in spite of its small size, the infant thyroid may be more susceptible than the adult thyroid to cancer induction by ionizing radiation."

No one denies that radio-isotopes produce damage when they are deposited in the human body. Controversy tends to centre around the "maximum permissible concentrations" (MPC's) that have been established for the quantities of various radioactive elements that the human body can be allowed to accumulate. There is nothing safe about an MPC. An MPC constitutes the amount of risk an official agency is prepared to inflict upon certain individuals and the general population in carrying out a nuclear-energy programme.

The U. S. Naval Radiological Laboratory points out that any degree of exposure to ionizing radiation produces "biological effects." "Since we don't know that these effects can be completely recovered from," observes the laboratory's report to the Special Subcommittee on Radiation, "we have to fall back on an arbitrary decision about how much we will put up with; i.e., what is 'acceptable' or 'permissible'—not a scientific finding, but an administrative decision."

Many scientists take a grim view of the "administrative decisions" that have established the permissible levels of strontium-90 for the general population. The MPC for strontium-90 is measured in strontium units (S.U.), formerly called "sunshine units." A single S.U. is one micromicrocurie (¼¼c) of strontium-90 per gram of calcium (a ¼¼c is equal to one millionth of a millionth of a curie; a curie essentially represents the amount of radioactivity associated with one gram of radium).

For a number of years, the maximum permissible concentration for strontium-90 was established by a definite although largely unofficial procedure. The MPC originated as a recommendation by the International Commission on Radiological Protection (I.C.R.P.), an advisory body made

up of scientists from all parts of the world. The International Commission's recommendation, in turn, was usually adopted by the National Committee on Radiation Protection (N.C.R.P.), the American affiliate of the I.C.R.P. Finally, the National Committee's recommendation was generally adopted by government agencies.

Until 1955, the recommendation of the International Commission dealt almost exclusively with problems of occupational exposure to radiation and radioactive isotopes. The problem of formulating an MPC for the general population was left in the hands of the commission's national affiliates and official agencies. This created a highly unsatisfactory situation. It made it possible for official agencies to grossly understate the hazards of nuclear weapons tests; they proceeded to evaluate all the dangers that fallout presented to the general population in terms of the large MPC established for workers in atomic plants.

"The maximum permissible level of strontium-90 in the human skeleton, accepted by the International Commission on Radiological Protection, corresponds to 1000 micromicrocuries per gramme of calcium [1,000 S.U.]," noted the British Medical Research Council in the mid-1950's. "But this is the maximum permissible level for adults in special occupations and is not suitable for application to the population as a whole or to children with their greater sensitivity to radiation and greater expectation of life."

To cope with this problem, the International Commission decided in 1955 that the prolonged exposure of a large population to radiation should not exceed one tenth of the maximum permissible levels adopted for occupational exposures. For all practical purposes, the commission's recommendation meant that the MPC for strontium-90 established for the general population should be reduced from 1,000 to 100 S.U.

It was not until a great deal of controversy had developed over the established MPC for strontium-90 that the National

Committee, followed with undisguised reluctance by the A.E.C., adopted the "one-tenth rule" recommended by the International Commission. It soon became evident that even this 90 per cent reduction was inadequate. Finally, the International Commission made a curious, perhaps contradictory decision: It raised the MPC for workers in atomic plants from 1,000 to 2,000 S.U. but recommended that the MPC for the general population be reduced to one thirtieth of the occupational level. Thus, in a circuitous, often confusing manner, the permissible level of strontium-90 for the general population has been reduced to 67 S.U., or one fifteenth of the MPC (1,000 S.U.) that was in effect in 1954.

The problem of establishing a "suitable" MPC, however, is still unsettled. Strontium-90 is not uniformly distributed in bone matter. The element, like radium, tends to form "hot spots," some of which may exceed the average skeletal distribution many times over. To estimate the damage which skeletal concentrations of strontium-90 are likely to produce, a factor, *N*, should be used to increase any average result based on a uniform distribution of the element in the bone—especially in the case of adults, who form these "hot spots" more readily than children.

Two Swedish investigators, A. Engström and R. Björnerstedt, who have pioneered in research on this problem, emphasize that *N* is not constant. The factor may vary from 6 to 60, depending upon individual metabolism and variations in the amount of strontium-90 that appear in food, air and water. An individual who acquired an average, presumably "safe," skeletal burden of 65 S.U., for example, might well have minute "hot spots" of nearly 4,000 S.U—enough to increase appreciably his chances of developing cancer.

"Meanwhile, what is the course of wisdom?" asks W. O. Caster, of the University of Minnesota. "Some claim that where there is honest doubt, public safety demands that the safety standards be adjusted to cover the worst possible contingency. But if one couples Dr. Engström's estimate of 100 strontium units as the maximum permissible level for radiation workers with the International Commission's suggestion that the permissible level for a population should be one thirtieth the occupational level, it would appear that the population limit should be only three strontium units. Some children have already passed this mark. The official agencies point out that, in the absence of proof that such a level is in any way deleterious, it would be the height of irresponsibility to raise a public alarm." The proof required by official agencies, however, might not be forthcoming for ten or fifteen years.

Whatever the analytic method employed, there is no doubt that concentrations of strontium-90 in human bones have been rising steadily over the past decade. The principal source of American data on the amount of radio-strontium entering the human body is an A.E.C.financed research project at the Lamont Geological Observatory of Columbia University under the direction of J. Laurence Kulp. Thus far, Kulp's research group has issued four reports, covering the years 1954-9. The data show that for the "Western culture" area (Europe and the United States), situated between 30 and 70 degrees north latitude, the average amount of skeletal strontium90 in adults increased from 0.07 S.U. in 1954 to 0.31 S.U. in 1959.

The amount of radio-strontium in the bones of children up to four years of age rose from 0.5 S.U. in 1955 to 2.3 S.U. in 1959, a total seven times the amount in adults. Although the body discriminates in favour of dietary calcium against strontium, the discriminatory factor varies with age. As adults no longer undergo skeletal growth, they add only about one fourth of the strontium they ingest to their bones; children, however, add about half.

The value of averaging the amount of strontium-90 in bones at a given collection point and then averaging the averages for an entire "culture area," is highly questionable. Jack Schubert has pointed out that "when dealing with a potentially harmful agent involving the world's population, it can be misleading to use average values of strontium-90 in the bones. It is important, especially in relation to the setting of permissible levels for a world population, that we have some basis for estimating the degree to which appreciable fractions of the population accumulate two and more times the average amount." Using more suitable mathematical methods than those employed by Kulp and his colleagues, Schubert shows that 28 per cent of the skeletal specimens analyzed in the 1957-8 Kulp report "have *three* or more times the average (geometric mean) amount of strontium-90 in their bones. Over 4 per cent will have seven or more times the average! These values are appreciably greater than those of Kulp's who used incorrect averages and an incorrect distribution curve and hence underestimated the fraction of the population which would exceed the average values."

Submerged in the avalanche of averages, estimates and statistical extrapolations are those communities that have been heavily irradiated by fallout debris. "Substantial areas in Nevada and Utah close to the bomb testing grounds had received ten roentgens of gamma radiation as early as 1955," Caster observes. "Some 40 communities had had average doses between one and eight roentgen units. Such doses are substantially above allowable levels for the general population—or for professional personnel for that matter."

In May 1953, for example, a sudden change in weather caused the fallout "cloud" from a Nevada test series to move over well-traveled and inhabited areas near Yucca Flats. According to an account of the episode by Paul Jacobs, fairly high levels of air contamination were recorded in St. George (population, 5,000) and Hurricane (population, 1,375). Hundreds of motor vehicles traveling the highways near the test site required decontamination, while residents of endangered communities (when they could be reached) were warned to remain indoors. At St. George, reports Jacobs, a "high degree of contamination continued for sixteen days after the shots."

Radioactive debris from the Nevada test explosions normally drifts in a northeasterly direction toward the grain and dairy states of the upper Midwest. It tends to settle out in areas where a large percentage of the American food supply is raised. As a result, high concentrations of strontium-90 have been found in food plants, especially cereals, grown in many parts of the Great Plains region. In 1957, for example, wheat samples collected from seven agricultural experiment stations in Minnesota registered concentrations of strontium-90 about 100 per cent higher than the current daily maximum permissible level (MPL) for humans over a lifetime. The largest number of S.U. found in one group of Minnesota samples was 200 per cent greater than the MPL.

Attempts have been made by highly responsible scientists to estimate the damage caused by the fallout debris produced to date. According to A. N. Sturtevant, professor of genetics at the California Institute of Technology, fallout will produce harmful genetic mutations

in 4,000 people in the first generation and in 40,000 people in generations to come. Although these figures represent Sturtevant's latest published estimates, they are calculated on the basis of the radioactive debris produced up to 1956. Since then, the total amount of fallout has increased about 70 per cent. On the basis of more up-to-date material, James Crow, president of the Genetics Society of America, estimates that 20,000 mutations will be inherited by the next generation as a result of fallout (his estimate is based on the assumption that the present world population will produce a total of two billion children). The number would reach 200,000 if nuclear testing were resumed and continued for thirty years at the 1954-8 rate.

If E. B. Lewis's "linear hypothesis" is correct, fallout may have contributed to the rising incidence of leukemia. According to Lewis's theory, the chances of acquiring leukemia increase in direct (or linear) proportion to increased irradiation. The hypothesis "predicts that constant exposure to even one sunshine unit [strontium unit] would be capable of producing about 5 to 10 cases of leukemia annually in the United States population." There seem to be very few scientists who deny that linearity exists in the middle dose range of radiation, but a great deal of controversy has arisen over Lewis's hypothesis that linearity exists at all levels of radiation, including the very low dose range involved in diagnostic X-rays and fallout. Yet there is a substantial amount of data to support his theory.

Aside from the basic evidence that Lewis presented when his views were first published, the linear hypothesis has gained strong support from the work of Alice Stewart and others on cancer in children. The pelvimetries studied by Stewart and her co-workers involved dosages of only 2 to 10 rads; nevertheless, this low level of whole-body radiation directed at the fetus was enough to double its chances of acquiring cancer.

If Lewis's hypothesis is also valid for other forms of radiation-induced cancer, fallout from nuclear weapons tests may be responsible for thousands of cases of bone cancer. By using a statistical approach developed by the British Atomic Scientists Association, it is possible to make a rough but reasonable estimate of the number of such cancers that would be produced by increases in the amount of strontium-go in human bones. A long-term average of only one strontium unit in the bones of every individual in the world may well cause 12,500 cases of bone cancer; as little as two strontium units may result in 25,000 cases. The figure increases proportionately as strontium-go becomes part of the human skeleton. It is sobering to note that if the world population were to acquire an average skeletal concentration of 20 strontium units—or one third of the current maximum permissible concentration—a quarter of a million people might be afflicted with bone cancer.

All of these estimates and hypotheses may be wrong, but thus far there are no convincing reasons for believing that they err on the side of pessimism. "The by-products of atomic fission are highly destructive to protoplasm," observe Ralph and Mildred Buchsbaum, "and in unique ways which present problems that were never posed by the other hazards man has introduced into his environment by his technology. The dangers are at present underestimated, rather than exaggerated, in the opinion of most biologists."

THE NUCLEAR AGE

If Nuclear Weapons Tests were ended permanently, we would still be confronted with far-reaching dangers created by the peacetime uses of nuclear energy. In contrast to the minuscule quantity of radioactive substances to which man had access before 1940, the nuclear-energy industry produces millions of curies of waste materials every year. The "high-level" wastes created by the reprocessing of partially spent nuclear fuels may contain as much as several thousand curies of radioactive substances per gallon.

More than 65 million gallons of these lethal materials are stored in giant underground tanks at three locations in the United States. So-called "intermediate" and "low-level" wastes are discharged directly into the ground. Radioactive substances in liquid form, for example, are piped into gravel-filled trenches and allowed to percolate into the soil. Over 2½ million curies of solid and liquid waste products have been disposed of in the ground.

The liquid wastes "work their way slowly into ground water," observes the A.E.C. in its 1959 *Annual Report,* "leaving all or part of their radioactivity held either chemically or physically in the soil." To put the key thought in this statement less obliquely: Part of the radioactivity reaches the ground water and passes into man's environment. Large quantities of "low-level" wastes have been discharged into rivers.

At Hanford, Washington, for example, about 2,000 curies of substances that have been made radioactive by neutrons in reactors are released daily into the Columbia River. After making a survey of the Priest Rapids-Paterson section of the river, a group of U. S. Public Health Service investigators pointed out that "values as high as 2.2 × 10 ì/gm. [.002 microcuries per gram] have been found in the muscle of white fish near Priest Rapids." The writers reported that other river organisms "had gross beta activity densities considerably higher than adult fish, but these organisms are not utilized to any extent by humans. Thus, their major significance is in transmitting this activity to other organisms in the food chain. Fish as well as ducks, geese and other animals, may consume these organisms." These higher animals, it should be noted, are consumed by man.

In addition to using rivers as a means of disposal, the A.E.C. has dumped 65,000 concrete-lined drums containing "low-level" radioactive garbage into the Atlantic and Pacific oceans. Owing to the corrosive action of sea water, the porosity of the concrete liners and the high water pressures at the floor of the ocean, the drums may last as little as ten years. Approximately 60,000 curies of radioactivity were disposed of in the oceans before the A.E.C., yielding to widespread public protest, declared a "moratorium" on this type of disposal in all Atlantic and many Pacific coastal waters pending a further study of how effectively the wastes are retained by their containers. The A.E.C., however, is by no means the worst contributor to radioactive pollution of the oceans. British authorities allow their Windscale installations to pump as much as 10,000 curies of short-lived radio-isotopes into the Irish Sea every month.

What happens to these wastes when they enter the environment of fresh- and salt-water marine life? Generally, ecological factors play a number of ugly tricks on man; it has been clearly established that radioactive materials in water tend to be concentrated by plankton, algae, mollusks and fish. "Available substances are rapidly taken up by the biota, never remaining long in the water to be diluted and washed away," observes Lauren R. Donaldson, of the University of Washington. "This is dramatically demonstrated following an atomic test in which radioactive materials are deposited in the water.

Within hours, the great bulk of these materials is to be found in the living organisms. Plankton and some of the algae, which are the key organisms in the food chain, may concentrate within themselves more than a thousand times the amount of radioactive substances found in the sea water. The herbivorous fish and invertebrates have lower concentrations of radio nuclides at any given time than do the plants on which they feed and progressing along the food chain to the carnivores the concentrations become lower and lower."

There have been cases, however, in which the concentration of radioactive elements was higher in fish and mollusks than in plants. In a study of fallout on a fouracre pond near Cincinnati, L. R. Setter and A. S. Goldin, of the U. S. Public

Health Service, found extreme variations in the concentration of radioactive isotopes among different species of water life.

In some algae the amount of radioactivity was only 20 times greater than that in the water, but bass had from 300 to 400 times more and snails from 1,000 to 3,000. Data supplied by Richard F. Foster, of the Hanford Laboratories, indicate that whereas algae in the Columbia River concentrate radiocesium by a factor of 1,000 to 5,000, fish concentrate the element by a factor of 5,000 to 10,000. In any case, as one organism is nourished by another, the radio-isotopes reappear to a greater or lesser degree in the bodies of higher species. The route may be direct and simple or long and complex, but eventually a portion of the radioactive garbage finds its way back to man.

It is rather alarming to speculate on the direction the nuclear age may yet take. According to the Committee on the Effects of Atomic Radiation on Oceanography and Fisheries of the National Academy of Sciences, "300 nuclear powered ships of all nations will be in service by 1975. These ships could then potentially release to the marine environment approximately 5000 curies per year from expansion water, some 3400 curies per year from leakage and some 9 × 10 [900,000] curies per year from the ion exchange beds." The total: 908,400 curies per year.

At the same time, new reactors on the land can be expected to create ever-mounting problems of waste disposal. It has been estimated that if nuclear power furnished 10 per cent of America's current electrical output (a distinct possibility in the next few decades), there would exist in the world about 150 billion curies of radioactive isotopes from this source alone, including 2.6 billion curies of strontium-go and 2.3 billion curies of cesium-137. These figures, let it be noted, do not include radioactive wastes that would be produced by nuclear power plants in other countries.

Methods for the permanent disposal of such highly lethal materials would have to be "foolproof." If any miscalculations were made and the wastes began to seep uncontrollably into man's environment, they would cause a staggering amount of damage.

The nuclear age could have profound and far-reaching effects upon existing ecological patterns. Until recently, biologists assumed that the sensitivity of living things to radiation increases with the level of their evolutionary complexity. For example, viruses absorb between 50,000 and 100,000 roentgens before half or more of their number are killed; spore-forming bacteria, between 20,000 and 50,000; the weevil, between 1,000 and 2,000 roentgens. But mammals such as the dog, the goat and man begin to succumb to whole-body doses of less than 500 roentgens. In reviewing the striking differences in radio-resistance between lower and higher forms of life, many biologists concluded that if man and other mammals could be protected from radiation, the rest of the biosphere would more or less take care of itself.

It seems unlikely, however, that this conclusion is correct. There appear to be critical periods in the development of many highly radio-resistant species during which they can be damaged or killed by fairly low levels of radiation. For example, it requires only 160 roentgens to kill half the eggs of the fruit fly during the early stages of cell division, whereas 100,000 roentgens are needed to achieve the same result with adults. The organism is 600 times more radio-resistant at one stage of its development than at another.

As plants and lower animals tend to concentrate relatively large quantities of radioactive isotopes in their cells, a level of environmental contamination "harmless" to man may prove to be very harmful to less-developed species. The use of the soil, atmosphere and bodies of water as dumping grounds for nuclear wastes could result in serious ecological imbalances that would eventually affect man.

"When radioactive isotopes are released into the environment," observes K. Z. Morgan, of the Oak Ridge Laboratory, "they usually enter into complex physical and biological cycles and into food chains and may exhibit unexpected movements or concentrations. It is through such mechanisms that dangerous long-lived radionuclides such as strontium-go and cesium-137 are able to enter our food sources. These ecological food chains are actually circular in nature or web-like and have a limited number of links. Serious consequences would follow if too many of these food links were broken."

Morgan emphasizes that the damage can be very subtle. Radiation may reduce the mobility of an animal or impair its sensory organs, such as the eyes. It may reduce the organism's fertility. "Also, if a small dose has any effect at all at any stage in the life cycle of an organism, then chronic radiation at this level can be more damaging than a single massive dose which comes at a time when only resistant stages are present. Finally stress and changes in mutation rates may be produced even when there is no immediately obvious effect on survival of irradiated individuals."

Nuclear energy is associated with such huge problems in human affairs that the subtle dangers it creates for the biosphere may seem to be negligible. The tendency to deal with the atom in terms of massive destruction and large-scale economic exploitation may lower our guard and lead to serious ecological disturbances. Although the wastes discharged into the environment contain, for the most part, short-lived radio-isotopes, the daily volume of effluent is immense and the flow continuous. The radio-isotopes are picked up, concentrated and circulated in increasing amounts by key organisms in the vicinity of nuclear reactors and processing establishments until many plants and animals become heavily contaminated.

The smaller number of long-lived radioisotopes are carried further into major food chains that involve millions of people. Before creating new centers of radioactivity and adding to existing problems of waste disposal, we would do wisely to ask whether all the hazards of exposure to low-level radiation have been explored and whether all the alternatives to nuclear power have been exhausted. "It is not too late at this point for us to reconsider old decisions and make new ones," observes Walter Schneir in an excellent article on radioactive wastes. "For the moment at least, a choice is available."

Chapter 4

Ethnomedicine

In this chapter we discuss some of the more salient approaches taken by anthropologists in their analyses of illness, healing and those who provide assistance when sickness strikes. We discuss an anthropological interest that has evolved from curiosity about exotic beliefs and customs related to health and illness into a robust, rapidly growing, anthropological specialization.

We will emphasize the evolving approach to attributions of sickness, the social use of sickness and healing for purposes of social control and the recruitment and training of healers. We hope to demonstrate how these studies have contributed to the development of general theory and methodology in sociocultural anthropology, in addition to the emergence of the subfield of ethnomedicine. The chapter concludes with recommendations for the kinds of research necessary for the continued healthy growth of ethnomedicine.

OVERVIEW OF ETHNOMEDICAL

Healing, shamanism and the relationship between illness and supernatural forces have captured the interest of ethnologists and the public from anthropology's earliest days. These early classics reported from the far corners of the earth seemingly bizarre notions of the causes of illness and diagnostic procedures that invoked supernatural spirits or machinating spouses or neighbors. Accounts of the recruitment of diviners and counter-witchcraft specialists to discover the causes of illness had strong appeal for turn-of-the-century readers, whose own beliefs about health had just been radically altered by the industrial and scientific revolutions. By the 1930s research in the origins and provenance of the cultural components of medical systems was a prominent dimension of American cultural anthropology. Since ideas and behaviour related to sickness and healing were considered a significant part of culture, efforts to reconstruct the processes of culture building included close study of the tools and other paraphernalia of healers. In this effort, the distribution of culture traits related to health and the control of sickness were mapped and analyzed by Clements, resulting in the identification of five major causes of disease in the nonindustrial world: sorcery, soul loss, breach of a taboo, intrusion by a disease object and intrusion by a spirit.

It was concluded that a society can be characterized by the disease cause most prominently reported for it (e.g., as a spirit-intrusion society or a soul-loss society). In addition, Clements sought to infer from the spatial distribution of these traits how long they had been a part of a particular culture under study.

Clements's work was roundly criticized by researchers who saw more promise in the configurationist approach being advanced by Ruth Benedict in her Patterns of Culture. A major argument, forcefully pursued, contended that a configurationist approach placed medicine in its cultural context: "What counts are not the forms but the place medicine occupies in the life of a tribe or people, the spirit which pervades its practice, the way in which it merges with other traits from different fields of experience."

His critique was a harbinger of a radical shift from a historical approach to research on health phenomena to an ahistorical, empirical orientation. The emerging functionalism viewed society as comprising interrelated parts, concepts of disease and its causes and the characteristic of healers being interdependent.

One of the earliest and, subsequently, most prominent ways in which ethnomedicine contributed to the development of theory and method in sociocultural anthropology was to show the functional integration of the components of health care institutions within society's cultural matrix, its social organization, or political system. The functional integration approach, together with what have become known as the cognitive and symbolic approaches, have become the dominant theoretical approaches to institutions of health care for the approximately half-century since Clements published his major work.

As anthropology became more systematic and research more sophisticated, ethnomedicine became one of the essential dimensions of culture investigated. As the community study method increased in popularity, especially in studies of Mexico and Guatemala, greater prominence was given to a society's conceptualizations of illness (the causes and cures), the role of healers and the relationship between concepts of disease and cosmology.

Links were identified between what had appeared to be bizarre health beliefs and practices and other aspects of the culture or social organization. These findings contributed to a more culturally relativistic attitude toward other peoples' health practices and understandings. Even the most exotic-appearing health beliefs and behaviors are made understandable in the cultural context in which they are found.

For example, the many productive studies of the attribution of sickness and death to witchcraft or sorcery identify them as prime sources for students seeking information on cultural interpretations of how and why sickness and death are visited on some members of a group but not others. The assumption that people become ill because they have been victimized by neighbors transformed into animals was rendered more accessible by the reports of Villa Rojas and others that in some societies, transgressions of social norms are believed to be sanctioned by illness visited on the transgressor.

Villa Rojas utilized the structural-functionalist approach to good advantage in analyzing the causes of sickness among Tzeltals of southern Mexico. That analysis appeared in the forefront of a number of important works in which associations were drawn between the social norms governing behaviour and the attribution of sickness to supernatural forces charged with maintenance of those norms. Sometimes those who use illness to punish a failure to follow the social norms are those holding official positions; sometimes it is the deceased or supernatural figures who play this role.

The connection is elegantly developed in a study of a small society of Native Americans, the Harney Valley Paiute. These Paiute have few formal mechanisms of social control (police, army, courts, or judges) but a welldeveloped fear of sorcery. Sickness attributed to sorcery is hypothesized as representing a societal sanction of socially unacceptable behaviour rather than interpersonal enmity.

This explanation seems plausible and does elucidate the fear of sorcery as a mechanism for maintenance of social order among the people of Harney Valley. But its even more vital contribution is to test the hypothesis that the attribution of illness to sorcery may contribute to the more successful functioning of some kinds of society with social organizations, like the Harney Valley Paiute, but not other kinds. As a result, we now know that in societies in which formal institutions of social control are absent or weak, sorcery attributions are more frequent. In societies in which institutions like a police force, courts, or an army are prominent, attributions of sickness to sorcery are less frequent.

In a subsequent examination of that hypothesis in connection with a Filipino Christian group under the hierarchical control of the modern Philippine Republic, sorcery allegations are found to be rampant. These allegations prove, however, to be associated with social discord that is not clearly assigned to a formal social control agency (e.g., police or army).

Competition for a lover, conflict between spouses, broken verbal agreements and arguments over ownership of land in which title is not unequivocally vested fall between the cracks, not being clearly in the domain of a provincial or federal court or clearly under familial control. In these cases, sickness is a form of punishment presumed exacted by means of sorcery. In other words, sorcery fills the void when responsibility for the resolution of conflict has not been clearly assigned.

Although anthropological interest in the attribution of sickness and other health phenomena dates from at least 1915, it was not until 1968 that the term ethnomedicine was applied.

In that year Hughes applied the term to "those beliefs and practices relating to disease which are the products of indigenous cultural development and are not explicitly derived from the conceptual framework of modern medicine." Subsequently ethnomedicine was applied more broadly to refer to "culturally oriented studies of illness," and it was argued that the concern of the ethnomedical investigator was to explain "an illness—its genesis, mechanism, descriptive features, treatment and resolution—as an event having cultural significance".

One year later, ethnomedicine was defined as the "study of how members of different cultures think about disease and organize themselves toward medical treatment and the social organization of treatment itself". More recently, Nichter described ethnomedical inquiry as the "study of how well-being and suffering are experienced bodily as well as socially, the multivocality of somatic communication and processes of healing as they are contextualized and directed toward the person, household, community and state, land and cosmos."

Consideration of ethnomedical matters in the holistic investigation of the cultural life of communities gave increasing prominence to the group's conceptualizations of illness, its causes and cures, the role of healers and the relationship between concepts of disease and cosmology. Additional studies more narrowly analyzed specific ethnomedical issues. That large amount of attention paid ethnomedical observations in the first half of this century contrasts with a view that anthropologists had paid little attention to medical matters in the first half of this century.

CRITIQUES OF THE ETHNOMEDICAL

Critiques of the rapidly growing field of ethnomedical studies draw attention to its peculiarly mentalistic orientation. Fabrega wrote: The implicit assumption adopted by the researcher is that he is dealing with a disorder that is either typically psychiatric or at least psychiatric-like. Excessive preoccupation with this dimension on the part of culturally oriented anthropologists has tended to obscure the influences that biological components have on [culturally defined] illnesses. Consequently, the potential of examining the reciprocal influences that psychocultural and biological factors have on instances of illness occurrence [as defined and categorized by subjects] has been missed.

Indeed, ethnomedical studies are often conducted in societies in which such killer diseases as infant diarrhea, pulmonary tuberculosis, "river blindness," and schistosomiasis are rampant with little, if any, attention to the local population's cultural response to these diseases. Instead, research fastens on concepts, prevention and curing of folk diseases or diseases with psychiatric implications.

As a consequence of this emphasis, the impact of ethnomedical ethnographies on cosmopolitan medicine and, in particular, the culturally relativistic analyses of health institutions and practices in diverse societies have been primarily in broadening and making more flexible the psychiatric categories of the Interna-tional Classification of Disease and the Diagnostic and Statistical Manual tional Classification of Disease and the Diagnostic and Statistical Manual. The growing number of cosmopolitan physicians engaged in studies of how illness in other ethnomedical systems is constructed and responded to by patients promises new efforts to examine the implications of comparative studies for the diagnosis and treatment of patients.

ETHNOMEDICAL SYSTEMS

The effort to contrast and oppose "other" kinds of medicine with the increasingly dominant allopathy as practiced by medical physicians represents a serious problem, long impeding comparative medical studies. The difficulty contributes to a befuddlement as to the appropriate way to refer to allopathy, much less how to refer to those "other" systems so as to avoid invidious comparisons. It has been essential to find a nonpejorative term with which to describe biomedicine—the ethnomedicine in which medical physicians are trained—so as to be able to compare and contrast that ethnomedical system with others without making a priori value judgments.

We will refer to the former henceforth as cosmopolitan medicine to distinguish it from other ethnomedicines. To do otherwise is to fall prey to absurdities. For example, to describe the practices of curing among American, Mexican, or German laypersons as not modern when they are widely practiced in the year this is written is foolish. Similarly problematic is to label as non-Western those diagnoses or healing procedures extensively practiced today in suburban New Jersey or in the neighborhoods of Detroit or Los Angeles.

To refer to allopathy as the uniquely university-based medicine is to fail to account for the tens of thousands of physicians in India or Sri Lanka trained in the many university medical schools in which Ayurvedic medicine is the subject of lectures, laboratory training and clinical preparation, to say nothing of the thousands of licensed doctors of osteopathic medicine produced in the past decade by American medical schools.

For the reasons just indicated we will henceforth refer to the medicine practiced by physicians trained in biomedicine as cosmopolitan medicine and other ethnomedicines will be referred to by the name of the group of which it forms part of the culture (e.g., Bontoc medicine, Chinantec medicine, Zulu medicine).

ETHNOMEDICAL PARADIGMS

Much of the research on ethnomedical systems has focused on the questions of how ethnomedical phenomena are classified, how illness experiences are given meaning by patients and healers and how ethnomedical knowledge influences health-seeking behaviour. A recent development has been an increasing interest in the efficacy of traditional healing methods.

Access as to how a group classifies ethnomedical phenomena has been achieved by studying the language it uses to describe illness symptoms. In one pioneering study of Subanun diagnosis of skin disease, it was assumed that cognitive structures underlying illness behaviour and decision making were implicit in symptoms, which were systematically elicited from informants with standardized questions. That approach was subsequently improved by making the questions asked even more specific. In a study of curers and curing in highland Chiapas, Mexico, Metzger and Williams formulated their questions in terms of the informants' own concepts and categories rather than their own. They discovered unforeseen subtle distinctions in Tzeltal illness categories, illuminating a classificatory system far more complex than previously considered.

Other researchers, however, argue that not all of a people's knowledge about health and disease can be accurately represented by unidimensional semantic representations. Using multidimensional scaling instruments, they show how members of different cultural groups cluster diseases on the basis of a number of attributes or features. Comparing English-speaking Americans' and Spanish-speaking Mexicans' disease classifications, these researchers found that for both subject groups, diseases were not organized conceptually on the basis of the features that formally defined them but on the basis of pragmatic dimensions, such as type of victim, consequence or impact of the disease and kind of remedy indicated. This conclusion marked a shift away from features that define disease to the way people perceive the impact of disease on their lives.

HUMORAL MEDICINE

Humoral medicine, one of the most thoroughly studied topics in ethnomedicine, remains a subject of controversy. The concept of opposing humoral qualities' affecting health is a prominent premise of Latin American and other ethno-medicines. According to this theory, health is a matter of balance between the opposites "hot" and "cold," "wet" and "dry". One orientation to the widespread humoral concepts among Amerindians is to attribute their origins to the pre-Christian eras of Greek and Arab history.

Proponents consider that this humoral theory of health and illness was introduced to indigenous America in the sixteenth century by the Spanish conquerors. Four humors—hot, cold, wet and dry—constituted this important system. A person's state of health was to be attributed to a state of balance between such opposite qualities as hot and cold and wet and dry. Reference to these humoral qualities appears again and again in ethnographic accounts of Spanish and Portuguese-speaking people's efforts to engage in preventive health and to cure themselves or others of disease.

Compelling data have, however, been marshalled to show that prior to the arrival in America of Europeans, the concept of opposite qualities played a prominent role in health care of indigenous American societies. Other, more radical criticisms question whether such understanding of opposite qualities was systematically distributed among these populations, much less systematically applied as measures of preventive and curative health.

Although no one disputes that the concepts of hot and cold are remarkably widespread in Latin America or that they undergird much of the diagnostic and healing practices of these populations, whether of Indian or non-Indian cultural background, our concern is the extent to which this critical system is systematically distributed and used within these groups. For instance, we know that in one Spanish-speaking barrio of Tucson, Arizona, "no woman under 30 could make such a distinction [between qualities of hot and cold] or seemed to be aware of this system of classification."

In Guatemala, although women recognize hot and cold conceptual categories and their application to nutritional and medicinal decisions are widespread, they are simply not systematically distributed and perhaps never were. On the other hand, the free recall task Weller used by which to make salient the respondents' own domains of illness may not have provided an adequate key to tap the women's concepts of humoral qualities. When the humoral categories of disease provided by a sample of urban Guatemalans are compared with those elicited from women who live in the countryside, "the disease terms that were categorized beyond the chance level by the rural women were not necessarily the same ones significantly categorized by the urban women, nor were they necessarily categorized in the same manner." Inasmuch as a reader might conclude that these women simply fail to order things or concepts, such as humoral qualities, into categories, it is important to realise that these same women did systematically agree in their assessments of the levels of "severity" and "contagion" associated with familiar disease names.

Among highland Chinantecs of Mexico, although the conceptualization and utilization of humoral categories is widespread, a more encompassing system cross-cuts that of hot and cold. This one is based on mechanical understandings of human physiological processes. In the Chinantecspeaking municipality of San Francisco, the physiological processes of reproduction are managed by the use of medicinal plants, some of which are of a hot quality, others cold.

However, rather than using these qualities according to some arcane logic, townswomen pragmatically select humorally hot or cold plants to set off bodily mechanisms (e.g., to cause the uterus to evacuate its contents—whether the intent is to facilitate labour, prevent a conception, or cause a miscarriage—or to retain them—to prevent miscarriage or to arrest excessive menstrual flow or spontaneous menstrual hemorrhaging).

This analysis is important because it interprets the use of hot and cold concepts as facilitators of an underlying system by means of which essential physiological mechanisms are regulated. Those results also provide reason to investigate similar efforts to manage physiological processes in other indigenous groups. However, in another Mexican group, there are "no indications anywhere in Zinacanteco curing that h'iloletik [shamans] or their patients think or act about illnesses on the basis of concepts that involve the body as a system composed of functionally interrelated parts and processes—with the possible exception of body temperature." These investigations raise the question why the results and conclusions of so many investigations of the same humoral categories of disease are in conflict. Several students of the problem have asked the same question. Logan suggests that the problems attaching to research in humoral qualities plague ethnomedicine in general:

Most ethnographic accounts of humoral medicine are descriptive, in that no specific hypotheses or set of relationships are central to the given research. It is of limited use simply to report that a given people classify certain items or conditions as hot and others as cold. For without explanation, that is, without relating the data to some fact of the group's culture, ecology, or biological adaptations, the data tell us little more than certain items and conditions are judged, by some informants, to be hot or cold.

It is possible that the questions researchers ask about humors are phrased ambiguously, leading to different answers to apparently similar questions. Unfortunately, this cannot be evaluated, inasmuch as readers are seldom provided information as to the questions asked. Neither are readers provided the criteria by which respondents are included in the sample, the sample size, or how the data were collected. In view of the fact that organizing hypotheses are not utilized and interviews or information as to the constitution or size of the sample are commonly unavailable, collection of a cumulative body of data has been problematic.

Furthermore, earlier studies mistakenly assumed homogeneous understandings of humoral categories, failing to account for intracultural variability or variation within individuals' accounts in the context of such life events as sickness, pregnancy and lactation. Until such methodological problems are resolved, it is too soon to conclude that ethnomedical domains are not systematically distributed within a social group.

THE MEANING OF ILLNESS

The influential concept of explanatory models of illness also recognizes the importance of context. Explanatory models are sets of beliefs or understandings that specify for an illness episode its cause, time and mode of onset of symptoms, pathophysiology, course of sickness and treatment. Explanatory models are "formed and employed to cope with a specific health problem and consequently they need to be analyzed in that concrete setting".

Although others have invoked the concept of an explanatory model of illness as a cultural construct, Kleinman makes it clear that his explanatory models are attributes of individuals, drawing upon general cultural knowledge but remaining at least partially idiosyncratic and situational. Kleinman's work and that of many others influenced by his approach, however, fail to specify in any detail the extent to which individual explanatory models are shaped by culture and the extent to which they are idiosyncratic formulations.

A major innovative effort to analyse this important issue is found in a study of the Canadian Ojibway. Combining in a single study the explanatory model and the cognitive approach to health understandings, Garro found an average of 78 percent of the personal explanations—explanatory models—reflect the shared Ojibway cultural model. Using this approach to understand Ojibway explanation of high blood pressure, we now know that individual Ojibway have explanatory models that draw from the group's shared cultural. knowledge about high blood pressure.

Another limitation of the explanatory model approach is that models focusing on designation and classification do not inform us about the links between illness and social context. In contrast, on the basis of fieldwork in the Iranian city of Maragheh, Good examined the complaint of heart distress in its social and cultural context.

Using information on the distribution of heart distress in a stratified sample of 750 persons from Maragheh and the surrounding area in conjunction with an explanatory model drawn from local traditions of Galenic-Islamic and sacred-Islamic medicine, he described the social and affective context of the experience in terms of a semantic illness network: the "words, situations, symptoms and feelings which are associated with an illness and give it meaning for the sufferer".

In his view, heart distress not only served as a vehicle for the expression of social stress but was also used instrumentally to bring about action to relieve it. Elsewhere, in another study about the ways in which a group speaks and thinks about sickness, it was observed that over the course of an illness episode, villagers change the labels assigned the signs and symptoms of the condition.

In these observations among the Ngbandi of Zaire, Bibeau concluded that to understand medical language, one must examine it in context. Bibeau's observation that the Ngbandi name for a disease changed from one context to another (e.g., when discussing the site at which the condition is located, or the extent to which it resembles an animal, or if it is assumed to represent a social sanction for inappropriate behaviour) led him to develop the idea of a network of names associated with a particular disease. Each name in the network refers to a different feature or characteristic of the disease, changing from one context to another. Bibeau discusses six principles that underlie the origins of the different labels used contextually, or what he refers to as "speech situations."

Like D'Andrade, Bibeau emphasizes the productive or generative capacity of language systems. Bibeau attempts to contextualize verbalized medical categories and forge links between them and the social settings in which they occur. Yet another effort to contextualize health knowledge focused on therapeutic narratives—"commentary on illness progression, curative actions and surrounding events"—that occur naturally among members of the lay therapy management group. Others have also written of how symbols of ill health serve multiple social purposes.

Whereas ethnomedicine's insistence on providing the cultural context in which an illness is analyzed has enriched our understanding of how cultures construct illness and has expanded the forms that illness can assume, that emphasis has inhibited cross-cultural comparison. A methodology has recently been proposed that might render certain types of cross-cultural comparison ethnographically valid and systematic.

When cultural processes are anchored to physiological

mechanisms, which are the same species wide, responses to those referents by different cultural groups can be more readily compared.

For example, comparative ethnographic descriptions of illness in infants in which diarrhea is a salient physiological process illustrate what might be done with this approach. Many ethnomedical studies describe chronic infantile diarrhea, accompanied by a depressed fontanel, eyes seemingly sunken in the face, apathy and sometimes vomiting. Cosmopolitan medicine refers to this congeries of symptoms as dehydration associated with diarrhea. Other cultural groups, such as the sixteenth-century Aztec; contemporary Hondurans, Peruvians and Brazilians; Mexican Americans of Texas; Shona and Ndebele of East Africa; and East Indians respond in diverse ways to the same signs and symptoms of biological dysfunction.

Some of these ethnomedical systems attribute the condition to the escape of the infant's vital force through the still-open fontanel. In others it is attributed to pressure or trauma, which depresses the fontanel, in turn causing the upper palate to block the oral passageway. Elsewhere the condition is attributed to the child's mother's breast feeding him or her too soon after she has been exposed to a woman who has recently experienced a miscarriage. In another system, biomedicine, the same condition is explained by intrusion of an enteric pathogen, causing the infant an infection, of which loose stools is a consequence, whereas in parts of Brazil it is a consequence of the evil eye.

It is striking that some ethnomedical systems explain the condition as dehydration, of which a depressed fontanel is a manifestation, while others understand the characteristic runny stool, fluid loss, sunken eyes and fitfulness to be caused by the depression of the fontanel.

ETHNOMEDICINE AND HEALTH-SEEKING BEHAVIOUR

Ethnomedical research has made a significant contribution to the understanding of how knowledge about illness influences health-seeking behaviour. But recent investigations have emphasized that verbalized illness categories, explanatory models and other knowledge are not predictive of consequent behaviour. Two studies, one in urban Nepal and the other in rural Mexico, provide strong evidence that illness beliefs are not good predictors of the health-seeking strategies of patients and their families.

Nepalese, who have access to alternative forms of treatment, behave according to two basic patterns: illness specific, in which they seek out different kinds of therapy for different disorders and multiple use, in which assistance is sought from a variety of different medical resources during a single episode of illness. When asked informants expressed a preference for an illness-specific strategy; when observed, their behaviour reflected a multiple-use strategy. The discrepancy may occur because therapeutic choices reflect the beliefs and preferences not only of the patient but of family and friends as well.

In rural Mexico, Young and Garro seized on an opportunity presented by two rural, Tarascan-speaking villages similar in medical beliefs but differing in degree of access to cosmopolitan medical facilities. In this quasi-experiment the researchers tested two competing explanations of why Indians fail to utilize cosmopolitan health services:

- Because it is too costly in time and money to gain access to those services
- Because their health understandings are incompatible with those which guide clinic services.

Those in the village with bus service to cosmopolitan health services proved to use them at approximately twice the rate of the other. Other studies seeking to weigh the influence of, respectively, a group's ethnomedical knowledge and the ease with which it can utilize cosmopolitan medical clinics have produced different results. Apparently, the differences from Young and Garro may be attributed to the fact that the more recent studies are of patients with the socially stigmatizing disease of leprosy. To avoid becoming known and stigmatized as lepers by their neighbors, these Nigerian and

Nepalese villagers bypassed local clinics, spending additional money and time to obtain care far from their home.

The linkage between medical knowledge and social context becomes considerably more complicated the more medically pluralistic a society is. For example, in Belize there are a variety of medical resources available to residents: cosmopolitan medicine in the form of a hospital, outpatient clinic and pharmacy; "bush medicine" consisting of traditional practitioners outside the cosmopolitan framework; Catholic or pentecostal spiritualist healers; and household lay knowledge.

Staiano argues that in a pluralistic setting, such as Belize, there are alternative systems of interpretation from which the patient and healer can select. Two case studies serve to illustrate the continuing process of negotiation that goes on as the patient seeks therapies and etiologies consistent with his or her understanding of an illness. In both cases, the patients and their families accept some aspects of the cosmopolitan health care system as presented to them by a government physician, but they supplement this with information gathered in consultation with traditional healers.

The dynamic process by which patients accept, reject and adapt information provided by health care providers is exemplified in research from Detroit. In this study of middle-class women of Detroit, all of whom are diagnosed as hypoglycemic, the women incorporate the physician-provided diagnosis, adapt it to their preconceived concepts of disease and utilize it to meet the needs and exigencies of their already established lifestyles.

HEALERS

From the earliest periods of anthropology, there has been a marked interest in the recruitment, training and personality of traditional healers. Indeed, interest in the varied ways in which cultures permit the evidencing of psychopathology extended to speculation that the behaviour of healers who cured their patients with dramatic ritual procedures was to be explained by emotional aberrance. One authority has even defined shamanic procedures in Siberia as evidence of mental illness: "the Siberian shaman... undoubtedly is psychopathic. Mental illness plays a great part in his life and behaviour." That view is no longer accepted.

Although many popular books and much of the medical literature make reference to "the traditional healer," it is abundantly clear that there is no such universal entity. In the following passages, some more accurate generalizations will portray traditional healers.

Traditional healers are recruited in several ways. One common way is by divine selection, in which an individual has a dream in which it is indicated that the dreamer is to undertake healing responsibilities. Another form of divine selection is the normative obligation of a person who must accept such a responsibility in return for his or her recovery from an acute or lifethreatening illness. An elected person who fails to become a healer is thought to be subject to divine sanctions, inclusive of serious illness and death.

In many other societies, in contrast, an individual who aspires to healing apprentices to an established practitioner. The apprentice learns through didactics as well as observation of the master's conduct. In such situations, the student usually compensates for the training by assisting the senior healer and the latter's family. Sometimes a neophyte seeks out a noted healer, paying a fee to be trained. Finally, there are some societies in which individuals who wish to embark on a curing career and possess the necessary self-confidence to engage in such risky behaviour initially test their abilities on household members and close family and subsequently expand their practice to nonrelatives.

There are several important differences between divine selection and formal training by means of apprenticeship. For one, people who are divinely selected or elect to learn by the seat of their pants are less subject to social control than those who undergo an apprenticeship or other formal training programme. Inasmuch as traditional healing often includes considerable management of supernatural forces, the fear that healers who gain such control through efforts not publicly bestowed and largely unsupervised may use such potent forces for "antisocial" as well as helpful purposes (e.g., sorcery or witchcraft) is realistic.

Some research has looked specifically at the relationship between healers' empirical knowledge and that of nonhealers. Anthropological attention has also focused on the personal characteristics that differentiate healers from nonhealers. For instance, in one ethnomedical study, shamans and nonshamans who reside in the Tzotzilspeaking town of Zinacantan were compared using both projective tests and data such as economic level, participation in the cargo system, levels of formal classroom education and acculturation to Spanish-speaking Ladino culture.

The shamans identified less than laypersons with the sociopolitically dominant Ladino culture and were more likely to perceive human figures and themes of interpersonal conflict in ink blots. The knowledge about sickness utilized by shamans in Zinancantan proved not substantially distinct from the comprehensions of laypersons, but the former are more willing to utilize that knowledge.

In another provocative study of Zinacantan healers, it was concluded that there was little to differentiate laypersons and specialized healers in their store of healing knowledge and little differentiation in knowledge between healers, but a considerable cognitive difference between healers and nonhealers. The healers were more likely than laypersons to impose their own sense of order on ambiguous stimuli, presented in the form of blurred photographs: less frequently saying "I don't know," giving greater variety of responses and tending to respond with their own categories rather than choices presented by the interviewer.

Elsewhere, among Tarascan-speakers in Mexico, whereas the medical knowledge of women curers and noncurers was substantially the same, the curers demonstrated far more consistency and agreement among themselves.

Although there is little in these reports to indicate how healers acquire their knowledge, a study from South Africa reports an unusual learning process: The extent to which knowledge is common among healers has received little attention, but in one interesting study, a Peruvian shaman was asked to comment on the symbolism employed in the altar of another local shaman: When the other was shown photographs of his rival's altar, he called it "fetishistic" and implied that it was founded on superstitions: With a disdain no less intense than Paz's, he asserted that Jose Paz does not really understand the illnesses he treats or the significance of the objects on his mesa. In Thailand, those who aspire to study with an established curer must often travel great distances to find a willing teacher.

Golomb described the social and economic factors leading to a geographically widespread network of traditional healers in Thailand. Consistent with his observation that within a particular geographic area there were considerable inconsistencies in diagnosis and recommended therapy among traditional healers, he noted that local healers rarely consulted with one another. Indeed, junior healers or patients visiting the home of a master healer usually traveled at least an hour by car from their own villages. Golomb argued that this lack of diagnostic consensus among local practitioners and the pattern of long-distance consultation were the result of intense competition among traditional healers for a limited patient pool. Competition from government-run clinics and the threat of competition from other healers made master healers loath to train local novices. This led aspirants to seek training from master healers who were geographically distant from their home villages. These long-distance networks were used for sharing knowledge and seeking consultation on difficult cases.

By seeking assistance from geographically distant experts, healers did not reveal their limitations to the local community, thus preserving their status in the regions they serve.The studies of how healers are recruited to serve crucial societal functions, how they acquire their knowledge and how they practice medicine retain a central position in ethnomedicine.

ASSESSING HEALING SUCCESSES

Another of anthropology's long-time interests, assessing

the success of traditional healing practices, has received new interest and reformulation. For example, Kleinman describes the assessment of therapeutic efficacy as the central problem in the cross-cultural study of healing. This issue is beleaguered by unsecured impressions, strong opinions and outright dismissiveness of traditional practices by some and undue romanticization by others.

Thus, it becomes vitally important to fashion ways to assess objectively the efficacy of traditional healing practices. As Etkin so usefully comments, "Plant use and other medical behaviors are effective if they effect or assist in producing the requisite, culturally defined outcomes". In Etkin's significant contribution to the study of ethnomedicine, she urges the importance of viewing traditional healing as a process containing several levels of success. The anthropologist can then determine the extent to which that process succeeds in returning the patient to normal well-being as well as assessing proximate successes achieved along the way.

Proximate successes are the appearance of hopeful culture-specific indications that the effort is having a desired effect (e.g., inducement of high body temperature, expulsion of phlegm, stoppage of bleeding). Proximate successes are distinct from those that are ultimate, that is, the full restoration of well-being. These studies of healers and their practices represent some of the most traditional interests in anthropology. Yet our inability to make generalizable statements or predictions as to the kinds of sociocultural environments that will produce one or more category of healer and our difficulties in evaluating the successes of traditional healing remain challenges still to be achieved.

RESEARCH FOR THE FUTURE

Future ethnomedical research might profitably focus on some of the following themes:

Symptoms of Sicknesses

It is only through tabulation and description of the symptoms reported by all patients suffering a particular sickness that we can hope to discover a consistent assemblage of indicators and identify the relationships among them. This kind of study can inform us whether individuals who complain of, for example, susto share more symptoms than a person who complains of susto and another who complains of mal de ojo.

We need to know as well to what people are responding when complaining of illness. For example, Lewis comments that although the Gnau of New Guinea complained of being ill, they based their reports not on the signs and symptoms but on other characteristics of the sufferer, "since in their view causes were not discernible from the clinical signs, exact description of these [signs] was not relevant".

Applicability of Ethnomedical Hypotheses

Such studies as Lieban's test among Filipinos of B. Whiting's linking of attribution of sickness to form of social organization and O'Nell and Selby test of Rubel's hypothesis of a relationship between failure to meet role expectations and heightened susceptibility to susto contribute to the comparative study of ethnomedical systems and eventually to a testable theory of illness and healing.

To sustain so valuable a direction, it is essential that the results of studies of ethnomedicine in one society be tested in other cultural groups to determine their general applicability.

The Interface between Biology and Culture

Ethnomedical researchers have shown a troubling "reluctance to explore the interface between biology and culture [which] derives from their belief that previous efforts to do so using the biomedical paradigm were failures because they forced rich, complex ethnographic data into artificial categories". However, by building on the anthropological tenet that human physiological processes are the same species wide, studies of cultural responses to such physiological processes as maturation, aging, gestation, pregnancy and delivery and disease offer the opportunity to maximize the equivalence of the processes to which cultural responses are being made.

The effort is, first, to design standardized comparable

units into which characteristics of, say, an illness, gestation period, or maturational changes can be placed and then to compare the response to them across cultural groups. Efforts to accomplish these goals have been few but noteworthy.

Effects of Healing Procedures

It is vitally important to discover the extent to which healing procedures directed at the improvement of social relationships have succeeded. Where the healing of an individual's health is a metaphor for the alleviation of social difficulties that threaten to rupture structural ties or social solidarity of his or her membership group, do participants feel or acknowledge that social ties are less threatened following the procedure?

In societies whose social structures are otherwise similar, do those that possess such metaphoric healing mechanisms experience less threat or disruption than those that lack them?

Healing Implications of Patient Support Groups

Although it is conventionally accepted that healing procedures that include patient support groups have better outcomes, the importance of this assumption urges empirical assessment. Does social support in a healing ceremony ensure better results and, if so, in what kinds of social organizations? Similarly, it is reasonable to question the extent to which the gender of either patient or healer influences the performance and success of treatment procedures.

Future research can provide a corrective to earlier tendencies to present a group's healing procedures as virtually uniform no matter what the gender, social status, or other important social characteristics of the participants.

Distribution of Particular Illnesses

Skewed distributions may be clues to underlying social or emotional mechanisms. Is a stipulated illness widespread or confined to one or several segments of the population? Do women and men suffer from it in equal measure? Does it affect persons without respect to their social and political status? Is it confined to a particular ethnic group or social class? For example, in ethnically pluralistic villages in Nicaragua, susceptibility to the illness grisi siknes distinguishes the Miskito-speaking residents from others. In Guatemala, discovery that two Cakchiquel villages with a common understanding of susto suffered significantly different attack rates contributed to insightful analysis of the socioeconomic stresses each was experiencing.

Distinctions among types of Healers

The choice of nonprofessional care when cosmopolitan physicians are available. Speculation that people prefer nonphysician healers because they share with them a paradigm of health and healing, or because lay healers take more time in treatment of patients than biomedical physicians, is unproductive. In several interesting studies of the patient-healer relationship.

It has been discovered that healers of whatever persuasion who are in great demand treat their patients impersonally and quickly. Furthermore, in one investigation, it is reported that these popular healers "generally fail to explain the etiology of the illness for which the patient is being treated, fail to share similar etiological beliefs with their patients and frequently fail to uphold the patient's claim to the sick role". Similarly Kleinman writes of "Chinese-style doctors" on Taiwan that "unless a patient asks they rarely explain about cause, pathophysiology, or course of illness. They may not even name the illness"

Chapter 5

Biomedicine

BIOMEDICINE AS A CULTURAL SYSTEM

In a series of classic articles, Doctors suggest that cultural systems can best be understood in terms of their capacity to express the nature of the world and to shape that world to their dimensions. Thus, for example, religion "formulates, by means of symbols, an image of a genuine order of the world."

This simultaneous shaping and expression produces a congruence between culture and experience that provides an "aura of factuality" within which cultural systems "make sense" and seem "uniquely real" to their participants. For our purposes, the crucial phrase here is "aura of factuality." The implication of Geertz's analysis is that cultural systems achieve a feeling of factuality, of realness, that is, in part or whole, a by-product of their symbolic forms.

In Western society biomedicine is generally believed to operate in a realm of "facts"; many people experience their most intimate contact with science through the biomedical description of the facts of bodily function and disease. This realm of bodily fact is often perceived to be quite separate from other cultural and social domains. "To a degree perhaps unique to segmented Western society, the participants of this ethnomedicine [biomedicine] emphatically distinguish their medicine from other aspects of institutions of their society. Illness is thought of as a 'natural' occurrence".

Given this assumption that nature and the body exist in a directly apprehendable realm of fact, the problem for a cultural analysis of biomedicine is the delineation of the "aura" in the "aura of factuality" that it promotes. The issue is not simply the description of biomedicine but the discovery of strategies that will make visible its nature as a cultural system. As Emily Martin points out, it takes a "jolt" to see the "contingent nature" of biomedical description. Several recent explorations of biomedicine undertake specific and deliberate strategies to provide this jolt by making visible the culture of biomedicine.

One strategy is historical contextualization; biomedicine is shown as the historically embedded product of particular cultural and social assumptions, thereby highlighting the "arbitrariness of institutions". Another strategy is to uncover, through analysis of metaphor and other forms of speech, ways in which social meaning is embedded in biomedical categories. Attending to the life worlds of clinicians is a third strategy; the daily practice of clinicians is revealing of biomedicine's theoretical and pragmatic foundations.

All of these forms of analysis aim to recover from the domain of the "natural" and the "given" those aspects of biomedicine that are cultural and constructed. Most historical discussions of biomedicine emphasize its origin in an elaboration of the Cartesian dichotomy between mind and body. Biomedical theory developed out of the possibility, following Rena Descartes, of a separation of the physical body from the mental and social. The body, as part of the natural world, becomes knowable as a bounded material entity; diseases similarly are physical entities occurring in specific locations within the body.

Robert Hahn and Arthur Kleinman describe the consequence: physical reductionism is a central tenet of biomedicine. This medicine also radically separates body from nonbody; the body is thought to be knowable and treatable in isolation. As Nancy Scheper-Hughes and Margaret Lock point out, even those who try to take an integrated perspective on illness "find themselves trapped by the Cartesian legacy. We lack a precise vocabulary with which to deal with mind-body-society interaction and so are left suspended in hyphens." This is not just a matter of vocabulary but of epistemology; biomedicine participates in deep-seated cultural assumptions about what it means to know the body. The particularity of this way of knowing the body can be seen in biomedical texts and practices that provide a mechanistic and desocialized imagery of bodily processes. For example, in a section of The Woman in the Body entitled "Science as a Cultural System," Martin examines the images of women's bodies found in medical textbooks and suggests that several metaphors of the body permeate their seemingly "scientific" (that is, in this context, neutral or valuefree) descriptions of physical processes. Thus, the processes of menstruation and menopause are described in terms of production and control.

The female reproductive system is geared to "production" and is organized as a hierarchical system of communication among hormones, cells and the brain. This imagery corresponds to that of our economic system. In menopause, "what is being described is the breakdown of a system of authority at every point in this system, functions 'fail' and 'falter.'

Follicles 'fail to muster strength' to reach ovulation. As functions fail, so do the members of the system decline". The key to this metaphor, Martin says, is functionlessness: "these images frighten us in part because in our stage of advanced capitalism, they are close to a reality we find difficult to see clearly: broken down hierarchy and organizational members who no longer play their designated parts". In these images, the "natural" functioning of the body is described in a way that fits a wider social view of women as defined by their reproductive function.

A similarly circular relationship between social and medical imagery can be seen in Rayna Rapp description of the process of genetic counseling. She points out that "statistics and medical terminology are genres of communication, not simply neutral vocabularies. Much of the scientific information that counselors want to convey is technical and invisible." The visual aids used by counselors, such as charts and graphs, have an effect in "shaping the perceptions of the client" and thus, for some clients, redefining what is known in terms congruent with the biomedical definition of the "natural."

The "codes, genres and assumptions construct the conversations genetic counselors may have with their patients", producing as natural a particular way of seeing the body and its reproductive life. A revealing account of the historical embeddedness of biomedical knowledge is provided by Michel Foucault. For Foucault, medicine is one of a number of related disciplines that have shaped the body as a vulnerable site for the articulation of social relationships.

In The Birth of the Clinic Foucault argues that modern medicine had its birth in the period around 1800 when medicine became clinically based and concerned with both the inside of the body and the control of the health of populations. Foucault's thought is complex and my discussion here limited, but two examples can perhaps give some idea of the sense in which he perceives that medicine both shapes and expresses its historical context.

Foucault describes the period around 1800 as one in which medicine shifted not from a less to a more accurate understanding of the body but from one kind of knowledge to another. Before 1800 Europe had a "medicine of species" that depended on classification; diseases were organized into families and species and related more to one another than to the body of the patient. Medicine after 1800 was dominated by what Foucault calls "the gaze," a new way of seeing that looked into the body and focused on what was individual and abnormal. Suddenly doctors were able to see and to describe what for centuries had been beneath the level of the visible. It was not so much that doctors suddenly opened their eyes; rather the old codes of knowledge had determined what was seen. A new way of seeing produced a new kind of knowledge: "clinical experience sees a new space opening up before it; the tangible space of the body the medicine of organs, sites, causes, a clinic wholly ordered in accordance with pathological anatomy". For Foucault the historical context and particularly its shaping of what is possible, of what can be seen, determines what at any time is considered to be true.

Practitioners of the early nineteenth century did not suddenly become better observers and therefore better able to discover the truth about the body; rather, there was a fundamental change in what constituted observation. This change brought about profound changes in medicine and these in turn shape the body we perceive. In this argument, the issue of shaping goes deeper than what is said. Foucault is interested in what can be said and in the mutual shaping of perception and possibility that gives rise to a particular medicine at a particular historical moment.

Foucault later extends this argument to show that in the nineteenth century, the body became an object of social control in a new sense. Minutely observed in clinics, prisons and hospitals, bodies could be made into docile instruments of and for the exercise of power. One tactic of discipline is the dossier—the collection of documents that locates, describes and accounts for each prisoner, patient, or child. As Foucault puts it, "The turning of real lives into writing functions as a procedure of objectification and subjection." Thus, for Foucault, neither "objective" description nor the case format in which such description is often framed constitutes value-neutral aspects of medicine.

Rather than functioning to delineate a reality that exists independent of its description, they are techniques for the shaping of reality that create patients as individuals susceptible to a particular kind of judgment. Thus, people are profoundly shaped by disciplinary mechanisms that permeate our society, with medicine primary among them. Issues of the relationship between mind and body, questions about what is knowable and integration into the discipline of institutional life are enacted in the daily practice of clinicians. An example of a study that explores the lived world of a practitioner is Robert Hahn "Portrait of an Internist".

Hahn portrays the symbolic world of a clinician; the internist uses and reflects on biomedicine's categories and his practice is revealing of how these categories exist in the larger culture. Hahn's strategy is to explore the interface between the personal and social that is provided by the world of work, showing the "goals, assumptions and uncertainties of medical logic". His internist enacts in work the production, of both self and society. The internist described in Hahn's portrait engages directly the questions of realism and nominalism inherent in biomedicine's Cartesian origins.

Thus, the internist "refers to his conception of the patient's problem, most often a physiological one, as 'a picture' a 'thing'"; sometimes "pictures" "make sense," and sometimes he "makes sense of" them. As Hahn points out, "If purported facts fail to make sense, the anomaly must inhere in the facts; but, if Barry (the internist) is unable to make sense of the facts, it may be that the difficulty lies in Barry's sense-making activity. These are respectively metaphors of realism and nominalism, of naturalism and constructionism".

Thus, this physician enacts, through work and in relation to the bodies of his patients, some of the fundamental issues embedded in the history of medicine itself. The assumption behind Geertz's definition of a cultural system is that "culture can be explained primarily in terms of itself". However, these examples suggest that the culture of biomedicine does not lend itself to explanation in terms of itself. One problem is the same as that of Hahn's internist in the passage quoted: the relationship between constructed and natural fact.

Hahn points out that social science observer in biomedical settings have often paid insufficient attention to its materiality. Biomedical practice depends on the assumption of an objectified nature subject to scientifically formulated "reality testing," and although, as Hahn points out, reality testing is fundamental to all healing traditions, we find our particular brand especially compelling. Thus, from the perspective of patients, practitioners, social scientists and laypeople in our society and despite much evidence of limitations or confusion, nature as it is understood by biomedicine demands to be taken seriously (that is, not questioned) in studies of biomedicine.

This paradox, usually not in evidence in studies of other medical systems—for example, most studies of Ayurveda do not generally consider its disease categories as descriptive of actual diseases but of socially constructed ones—means that the categories of the culture under study are also the categories used to study it. A second difficulty arises not so much in connection with factuality as with its aura. The closed circle of belief and expression suggested by the notion of cultural system appears flawed, even fragile, in several of these accounts.

This may result in part from the way illness itself threatens the cultural order with chaos and loss of meaning and thus "calls into question particular socio-cultural resolutions" of the dilemmas of human existence. Paradox and doubt may be intrinsic to the experience of the body; "physical form generates, from its own internal contradictions, the potential basis for critical awareness". In addition, however, biomedicine participates in a cultural separation of mind and body, nature and culture, in ways that may produce a sense of dissonance expressed in increasing criticism and doubt.

Martin, for example, found that women she interviewed expressed diverse images of their bodily processes, contradicting and resisting biomedical formulations. Similarly, Rapp's work suggests a complex interplay between social context and the expression of medical "information," with some counseling recipients unwilling to accept the language of risk in which advice was proffered and with that language itself constantly modified in interaction. Thus, as Jean Comaroff puts it, "there has been an awareness that 'factual' knowledge might imply social values, that medicine has bequeathed us powerful metaphors along with its 'natural' truths and that these might reinforce the deep-seated paradoxes raised by illness".

The examples given here suggest that critical perspectives tend to emerge out of the cultural analysis of biomedicine. Western biomedicine and medical anthropology are intimately connected. Many medical anthropologists work in biomedical settings or study problems that have been defined in biomedical terms. Medical anthropologists also study biomedicine itself, exploring the ways in which it is socially, culturally and historically constructed and showing how its perspectives influence the lives of its patients.

In addition, most medical anthropologists are members of societies in which biomedicine provides the dominant forms of explanation and treatment for illness and are thus participants in as well as observers of the culture of biomedicine. In this chapter I explore some of the implications and paradoxes of this relationship. My focus is on the ways medical anthropologists and others in related fields (mainly history and sociology) approach biomedicine as an object of study.

My emphasis is on biomedicine as it is understood by these writers; discussing the diversity, internal complexity and changing conditions of current biomedical practice in the United States is beyond my scope here. Recently a good deal of discussion and controversy has arisen within medical anthropology about its relationship to biomedicine. Often the issue is phrased as a difference between "clinically applied" and "critical" medical anthropology. Clinically applied medical anthropology has been described as "serving to clarify specific issues in health maintenance and response to sickness". Its orientation is the application of anthropological perspectives to particular clinical situations and problems.

Critical medical anthropology, on the other hand, defines itself in terms of a concern with the macrolevel of political and economic forces that shape medicine and determine the nature and extent of its interventions. Margaret Lock describes the critical approach as one that pays attention to "macro-structural questions, the role of power in social life and the way in which biomedicine is culturally constructed". Biomedical theory and practice is problematic not simply when it fails to address cultural and social issues involved in individual patient care but because of its embeddedness and (often) sustaining role in dominant political and economic systems.

The precise nature of the division between clinically applied and critical medical anthropology is by no means a matter of agreement among medical anthropologists and there are numerous variations on these definitions. Morsy points out that critical analysis is common in other disciplines and objects to using a label that sets it apart as special. On the other hand, M. Singer, Lani Davison and Gina Gerdis make a case for separating "critical" analyses that "explicate culture in non-cultural terms" from "culturalist" approaches that avoid economic or political forms of explanation.

BRACKETING BIOMEDICINE

One solution to the problem posed by medicine's grounding in "fact" is to segregate biomedical and social science ways of knowing. Most of clinically applied anthroplogy and much research in medical anthroplogy as a whole, is based on a bracketing of biomedical expertise as referring to areas of knowledge not within the purview of the anthropologist.

This bracketing is the basis for the well-known distinction between disease and illness proposed by Leon Eisenberg and Arthur Kleinman. This distinction is created by dividing up the field of "sickness" into a domain of disease, considered to be pathology as biomedically defined and illness, which encompasses the cultural meaning and social relationships experienced by the patient.

Allan Young sums it up thus: "Disease refers to abnormalities in the structure and/or function of organs, pathological states whether or not they are culturally recognized." This is the "arena of the biomedical model." Illness, on the other hand, "refers to a person's perceptions and experiences of certain socially disvalued states including, but not limited to, disease". Thus illness includes the experiences and beliefs of individuals; disease is what biomedicine discovers "in" the person regardless of his or her (personal or cultural) awareness.

The disease-illness distinction has provided the basis for much work in medical anthropology on the explanatory models and semantic illness networks of patients and, to some extent, of practitioners. These studies set aside the disease half of the distinction and concentrate on understanding the illness experiences and behaviour of individuals and cultural groups. By "setting aside," I do not mean that disease itself is not considered problematic for those who experience it but that the definition of disease—its status as a real, natural phenomenon-is considered nonproblematic. This has allowed medical anthropologists to study culture (beliefs, issues of meaning, experience of illness) in medical settings without dealing with questions of the cultural construction of medicine itself.

It also allows for the defining of research problems (for example, the study of groups of patients suffering from a particular disease or the study of the relationship between cultural and physical aspects of causation in a particular disorder) in ways that are relevant to the social context supporting the research. As Noel Chrisman and Thomas Maretzki say, "In our research, anthropologists have explicitly or implicitly drawn on clinical medicine as the standard for judging the 'real' world of sickness".

One consequence is that medical anthropologists have been able to do research and teaching in medical settings, finding ways to incorporate anthropology into practice while respecting the orientation and commitments of clinicians. For the anthropologist who is, as Chrisman and Maretzki describe, bicultural in anthropology and medicine, the ideal is a translation of perspectives, enabling clinicians to make use of anthropological insights. Often these insights have to do with negotiation among perspectives; at other times they have to do with patient advocacy or with the clarification of ways that the biomedical perspective influences the cultural interpretations of patients.

On the other hand, the disease-illness distinction is a variant of the mindbody and culture-nature dichotomies. By using it to separate natural facts from cultural constructions, medical anthropology runs the risk of taking on characteristics of biomedicine itself. Instead of offering a perspective that comes from a position of stranger, the anthropologist may be a kissing cousin in disguise. For example, the emphasis on case studies reproduces in anthropology the individual-centered and "objective" approach of the medical case study. Similarly, the use of scientific language to describe disease reproduces the position "from the outside looking over or into a space" that is fundamental to the medical gaze. The anthropologist is also influenced by the premise of biomedicine that "it is the medicine, real medicine; only other ethnomedicines are specially denominated, 'osteopathic medicine,' 'Chinese medicine'".

In both biomedical settings and the study of other kinds of medicine, it is hard to avoid the assumption that what needs to be explained are the "alternatives," the "other" perspectives, the "misunderstandings" or "misuses" of biomedicine rather than biomedicine itself.

An interesting recent development is that as biomedicine expands its definitions of physical disorder, incorporating problems with recognizably large social components (as in, for example, alcoholism and posttraumatic stress disorder), the position of the anthropologist becomes problematic. These conditions, with their roots in problematic social environments, seem to be ripe for anthropological analysis and understanding. However, attempts to bring social and cultural considerations to bear on biological phenomena tend to participate, often unwittingly, in a process of naturalization that turns them into things comparable to diseases. The bringing of chronic or behavioral conditions into the domain of biomedical treatment (the very thing that brings them to the attention of the biomedically based medical anthropologist) tends to result in their naturalization and "reinterpretation as events requiring medical intervention."

Thus, the more they are translated into the reified, concrete terminology of "disorders," the less room there is for the anthropologist's perspective on the cultural shaping of both the symptoms and their interpretation. As Young has shown for posttraumatic stress disorder, the production of "knowledge" about such disorders is itself a cultural process.

CRITICAL PERSPECTIVES

Much work in anthropology has explored the positive aspects of cultural systems in providing and sustaining meaning in human social life. But there is another perspective from which the congruence between the shaping and expressive aspects of culture can be seen as perverse. Religion, for example, appears in this view as an "opiate," preventing people from recognizing the truth of their situation. Medicine, in its powerful mediation of human physical and emotional frailty, can similarly be understood in terms of its relationship to a larger social (political and economic) system in which it serves to conceal sources of injustice and suffering. From this point of view, medicine cannot be described apart from the relations of power that constitute its social context. As Howard Waitzkin puts it: "Major problems in medicine are also problems of society; the health system is so intimately tied to the broader society that attempts to study one without the other are misleading. Difficulties in health and medical care emerge from social contradictions and rarely can be separated from those contradictions".

There are two aspects to this relationship. One is that health problems themselves may be socially caused, creating what Waitzkin calls the "second sickness". The other, related, aspect is that medicine may function to conceal the social origins of sickness and to suppress the possibility of protest.

When biomedicine is seen in this light, clinical knowledge itself becomes problematic; its connections to the larger system mean that it "cannot be either evaluated or transformed in any simple, decontextualized manner".

Nor can it be seen merely as a "web of significance" approachable through understanding; it must also (or perhaps, instead) be considered as a "web of mystification". Critical analyses of biomedicine are attempts at demystification. One strategy aims to uncover the incidence and causes of the "second sickness" by exploring ways in which medical care fails to reach, recognize, or correct socially created problems. Many analyses stress the relationship between capitalist production (and the profit motive inherent in it) and the failure to protect workers and others from its effects.

Others focus on the maldistribution of medical care and the effects on the health of populations created by the dominance of complex technology. A second strategy aims to uncover how biomedicine mystifies sickness through its participation in the nature-culture dichotomy. Medicine, because of its bias toward the uncovering of natural facts, represents the body in ways that are powerfully suggestive of a natural reality separate from the social. The effect, if not the intention, is to make the social invisible and to place sickness, as a natural process or entity, inside the individual.

Martin's point in her argument about menopause is that the "shriveling" of the ovaries is a metaphor that rests on and reinforces the social representation of the "shriveling" of production in the older woman. Because medicine has clothed the social representation in scientific language, it is difficult to discover its origins. Similarly, Michael Taussig describes the way a hospitalized patient is convinced of her own helplessness in the face of disease.

She minimizes her own strength because she has been taught to rely on experts who function to invalidate her intuitive understanding of the social origin of her problem. Her disease is treated as a thing, part of a natural world separate from the social world that oppresses her. Thus Taussig considers medicine to express a hidden ideology, one that reifies the social and separates it into a natural domain where it cannot be understood for what it is.

By placing the body and bodily experience in the realm of nature, biomedicine conceals both the social causes of sickness and the social embeddedness of the experience of sickness. Thus, for example, the diagnostic category of premenstrual syndrome (PMS) creates a "disorder" that may serve to obscure the social relations that are the context of women's suffering. Similarly, the processes of childbirth and dying may be isolated from their social contexts and treated in largely technical terms that prevent those involved from taking care of themselves and each other.

Recent cross-cultural and historical studies suggest that these tendencies toward reification and mystification are widely associated with biomedical practice. Lock's work on school refusal and on menopause in Japan shows that Japanese biomedicine similarly describes social problems as "syndromes" to be treated. In northeast Brazil, medical treatment, especially in the form of tranquilizers, serves to conceal the economic and social origin of starvation.

An example from the history of psychiatry comes from Andrew Scull, who shows that asylums in nineteenth-century England had the effect of isolating and controlling those in the population who could not survive under the conditions of early industrialization. Asylums maintained a distinction between the mad and the able-bodied, who could not be given relief for fear of undermining their value as surplus labour. Medical definitions of insanity contributed to and perpetuated the separation of "useless" from "useful" individuals. Scull sees the current move toward deinstitutionalization to be similarly motivated by economic policy; welfare and disability payments make it cheaper for the state to maintain disabled people outside asylums.

Other areas of medicine have also been seen as fostering dependence in order to conceal and support class and gender interests. E. Richard Brown, for example, shows that late-nineteenth-century capitalism in the United States deliberately fostered biomedical definitions of problems that might otherwise have been seen as related to industrial development. The notion of the body as a mechanism that could be repaired corresponded in important ways to factory production. Similarly, nineteenth-century medical theories about the fragility and emotionality of women served to bolster male dominance and the creation of the home as a domain separate from the workplace.

These analyses regard biomedicine's aura of factuality as precisely its source of power. Medicine can describe events in a value-neutral language that makes them appear to be part of the natural world and thus neutralizes what are, in reality, social problems. In the nineteenth century, villagers whose ability to support aging relatives had been undermined by social change were convinced that asylum care was provided by "experts" (doctors) and thus superior to their own; women who rebelled against restrictive conditions could be persuaded that bed rest was the only remedy for their restless female organs. Similarly, today, Brazilian peasants believe tranquilizers to be "medicine" for starvation and women angry over the unfair distribution of domestic work regard their anger as a "symptom" of PMS. For some writers this analysis of the embeddedness of biomedical categories in social life (and their tendency to perpetuate sickness-causing aspects of social life) is not enough. Additionally, it is important to recognize the ways in which biomedicine also gives rise to resistance. Martin attempts to make visible, through the analysis of women's speech, the way ordinary women resist the biomedical description of women's bodily life. For example, women may refuse to go to the hospital for childbirth, or they create original metaphors to describe bodily processes.

Brigitte Jordan, in an analysis of the medical "training" given to Maya midwives, shows that the midwives ignore much of what is presented to them and instead use medical supplies (masks, birth control pills) as props and symbols. They are resistant to changes in their way of delivering babies, preferring their own situated knowledge. Foucault suggests that this kind of "subjugated," situated knowledge, arising out of practice at a local level, forms the basis for a potential resistance to biomedical domination; however, he refuses to speculate about the ultimate shape that any change might take, insisting that while we can critique our system, we cannot be programmematic in our approach to change.

Those who emphasize the misuse of medicine are more prescriptive. If the problem is the creation of sickness under capitalism and the maldistribution and misappropriation of biomedicine, then the solution does not lie so much with changes in biomedicine itself or with pockets of resistance among patients or practitioners as in larger-scale changes in the system.

Hans Baer, Merrill Singer and John Johnsen issue this challenge: "Attention to the influence of classinterests as well as to the workings of power in large-scale organizations is vital for a truly critical medical anthropology. An approach that is sensitive to these issues will not cater to the furtherance of 'medical cultural hegemony' of the capitalist world system, but will help create a new medical system". Criticism of biomedicine—regardless of whether the stress is on discovering resistance or creating a new system—often seems to involve a paradox. On the one hand, biomedicine as part of society (the "medical establishment") is seen as failing to serve the real best interests of that society. On the other hand, the techniques of biomedicine (its science) are seen as one means for discovering these real best interests.

In some instances, biomedical categories themselves are employed to critique the use of biomedicine. For instance, Nancy ScheperHughes uses biomedical definitions of starvation to challenge the misuse of biomedicine to conceal it. This sidesteps the question, raised by those who consistently question biomedical categories, as to whether the science of biomedicine itself does not contain intrinsic assumptions about society and about the nature of reality that are, at best, disempowering and, at worst, harmful to body and society.

As an example of the complexity of this problem, consider Jordan's account of the training of Mayan midwives. Jordan suggests that these midwives are competent in their own right, rarely losing a mother or baby; she also suggests a few areas in which their management of labour and delivery is questionable by modern obstetrical standards. Is there a way to take what is "good" from biomedicine and incorporate it into their practice? Who should decide what that usefulness or relevance is, especially as medical standards themselves change rapidly?

Is it not possible that a few seemingly benign changes might undermine the midwives' entire practice? On the other hand, can Jordan, who knows, for example, that encouraging pushing too soon may damage the mother or baby, simply consider this aspect of the midwives' practice a part of their "culture," thereby refusing to acknowledge the possible benefits of medical training? In a situation like this, it becomes clear that we are torn between our own belief that the body can be considered part of the natural world, with at least part of its truth discoverable by biomedicine and our (often also strong) belief that biomedical intervention can be either oppressive or outright wrong.

Chapter 6

Essential Organic Substance

ALCOHOL AND DRUG

FUNCTIONAL DRUG: CANNABIS

Cannabis is a widespread drug of ancient vintage, usually classified as a hallucinogen based on its effects on mood and perception. Anthropologists have studied its use and its role in social structure in a number of cultures. Jamaica, for example, has a high rate of regular users of cannabis, making it an excellent place to study the issues that unfold around the use and abuse of this drug, known to Jamaicans as ganja.

In Jamaica, ganja use is integrally linked to all aspects of working-class social structure: cultivation, cash crops, marketing, economics, consumer-cultivator-dealer networks; interclass relationships and processes of avoidance or cooperation; parent-child, peer and mate relationships; folk medicine; folk religious doctrines; gossip sanctions; personality and culture; interclass stereotypes; legal and church sanctions; perceived requisites of behavioral changes for social mobility; and adaptive strategies.

Ganja figures strongly in the economic realm. On the lowest rung of the Jamaican socioeconomic ladder, poor families with few marketable skills make their living however they can, often engaging in several diverse economic activities. One of these activities may be the cultivation of cannabis. Vera Rubin and Lambros Comitas point out that vendors of cannabis typically lead a stable family life and are otherwise law abiding and conservative.

Working-class Jamaicans often believe that use of ganja makes work go more pleasantly and allows them to work harder; however, the middle and upper classes, who employ the lower classes and supervise their work, believe the drug is detrimental to work performance. Melanie Dreher has investigated the interplay of these differing views in the setting of a Jamaican sugar estate. Three farms were contrasted, all having different proportions of cannabis smokers to nonsmokers. Dreher examined productivity figures by categories of smokers and nonsmokers and found no significant differences in the work performance of smokers and nonsmokers. The results support neither the views of the workers nor the managers.

Dreher's study illustrates the increasing use of quantitative data in anthropological work on drug use, but it also points to the levels of subtlety and sophistication that can be added to research by the inclusion of qualitative material. This dimension of research is one of anthropology's strongest assets; it allows for deeper and more accurate understanding of complex questions and helps restrain impulses to hasty generalization.

Dreher has also studied the use of cannabis among Jamaican women. She examined the different patterns of ganja use among women in two similar Jamaican villages, where women in one village seemed more inclined to smoke cannabis than in the other. At the time of this study, smoking was contrary to norms for Jamaican women, although they routinely made cannabis teas for medicinal purposes. In the village with the higher rate of smokers, women were found to have more economic opportunity and thus more independence.

Women in the village with fewer smokers found it adaptive to conform to the norms, so that they would not alienate men who were potential husbands and sources of support. By taking a cross-cultural look at what is considered to be a growing problem in the United States, Dreher has undertaken applied research to shed light on an important policy question.

A common conceptual framework for the behaviour of drug users whose conduct is outside society's usual limits is the socialpsychological notion of deviance. Dreher argues that the behaviour of the women she studied is better understood anthropologically in terms of intracultural variation. Again, we see a contextual focus, a perspective that, instead of looking solely at individual action, allows for the connecting of individual action to sociocultural structure. Currently more tolerance for ganja smoking has developed among lower-class Jamaican women.

Use of the substance fits into a constellation of personal characteristics to which the term "roots daughter" has been given. Connected to the ideas generated by the Rastafarian religion about what is "natural" and African, a "roots daughter" is dignified, independent and intelligent. Findings of this later research indicate changing attitudes and thereby remind us of culture's dynamic nature.

Anthropological research often probes the role of ritual in human action, as in the case of the Colombian work of William Partridge. He found that smoking of cannabis was associated with work life through the action of ritual. An important part of worker comradeship is sharing cannabis when one can afford it; those who have it share it with whomever may not have it at that time.

Workers who use but do not provide cannabis for sharing at work breaks are considered undependable and isolate themselves from the social networks from which work gangs are developed in the agricultural economy. In this fashion, meaning is shifted from one social context to another. In a domain similar to ritual, Rubin has looked at the first experience of smokers as a rite of passage, an experience that strongly influences whether boys become regular smokers. If their first experience is a good one, they will probably become regular smokers; if not, they tend not to become users.

The practice of learning about the cultural intricacies of studied groups makes the anthropological enterprise often more time-consuming than other research approaches. In the Costa Rican work of J. Bryan Page, for example, a command of "proper" Spanish was not adequate to the task of following conversations in the argot of the drug culture on the street. The arcane vernacular was found to be useful to the speakers; one use was the concealment of illegal economic activity, economic activity again motivated by need.

Research in Costa Rica involved life histories, participant observation and ethnographic interviews and subdivided smokers into different categories whose members found their smoking experiences to be shaped considerably by conditioned expectations and socioeconomic factors. Users who enjoyed economic stability and good social support found the smoking of cannabis to enhance activity; those users with less secure lives did not have universally pleasing experiences.

In a keynote address to the Alcohol and Drug Study Group, Dreher documented several theoretical and methodological contributions from anthropology, especially as distinct from the research forthcoming from sociology and social psychology. Not surprisingly, the use of cannabis in modern society is viewed by anthropologists as a "sociocultural phenomenon rather than an individual characteristic". The holistic and comparative approach, the use of society or community as the unit of analysis, the reliance upon ethnohistorical and ethnographic studies and the combination of qualitative and quantitative methods make anthropological research on cannabis unique.

The cannabis question is obviously a loaded issue in American society at the present and therefore it is not surprising that there is resistance to accepting the results of these "controversial" anthropological findings. However, the same skepticism regarding these conclusions about the potential functional role of ganja use in certain cultural contexts is remarkably familiar to some of the resistance alcohol researchers in anthropology have encountered when they call into question certain widespread assumptions about the etiology and diagnosis of alcoholism.

The question of where to draw the line between acceptable usage—if usage of a particular drug is thought ever to be acceptable—and dysfunctional-pathological usage is extremely complicated. This message is probably one of the most important ones that anthropology offers, especially on controversial drugs such as cannabis.

SOCIOCULTURAL CONTEXT OF ALCOHOL USE

In 1940 Ruth Bunzel pioneered the way for anthropological research on alcohol through the publication of the results of a controlled comparison of the role of drinking in two different Central American societies where she had conducted in-depth ethnographic fieldwork. She had not set out to study drinking or alcoholism specifically. Applying psychoanalytic concepts, Bunzel connected drinking behaviour to its wider sociocultural context.

By identifying positive functions within these drinking patterns as well as explicating drunken behaviour, Bunzel found that drinking seemed to help lubricate social relations in the village of Chamula in Mexico. In comparison, in the Guatemalan village of Chichicas- tenango, alcohol provided a release from anxieties related to a stressful environment.

In the late 1950s Dwight B. Heath took up the gauntlet of this genre of research in his study of culture change following the Bolivian revolution of 1953. One area of investigation was change in drinking behaviour as an indication of shifting interethnic social relations between the peasants and the mestizos: "drinking is a useful index, being both highly visible and an integral part of the etiquette of relations between men". With this work and an earlier study among the Navajo Indians, Heath began the first sustained effort in anthropology to advance alcohol studies.

The number of ethnographies concerned with drinking practices and beliefs has proliferated since the 1950s. In Peru alone, three publications during the late 1960s and early 1970s focus on alcohol. In one example, Ozzie Simmons offers a description of the sociocultural integration of alcohol use within Lunahuana, a Spanish-speaking coastal village in Peru: "the meshing of drinking with a configuration of culture and social structure [gives] alcohol positive symbolic and functional roles" within the society.

Similarly, Paul Doughty found that "the use of alcoholic drinks was highly patterned and integral to normal social interaction" within the mestizo community of Huaylas in highland Peru. Allan Holmberg summarizes that for the agricultural peasant village of Viru on the north coast of Peru, "traditional patterns of drinking are such an integral part of the value structure of Viru that they are not likely to change in the near future".

A more recent example of basic ethnographic research out of which alcohol data developed but was not the central theme or intent of the research is Ndolamb Ngokwey's fieldwork among the Lele of Kasai in the Republic of Zaire in which he analyzes the drinking of palm wine: when it is drunk and within what contexts, types of wine drunk, drinking manners and its connection with health and illness. By concluding that "Lele rules and practices concerning palm wine reproduce cultural values, notions and categories", Ngokwey clearly articulates the sociocultural integration theme. He also raises another issue, which runs through much of the literature on alcohol (as well as other drugs): gender differences in consumption.

Gerald Mars, an anthropologist focusing on occupational issues, is another recent researcher whose studies uncovered distinctive patterns in drinking styles. Among longshoremen in Newfoundland, Canada, two distinct groups emerged: the "regular men" and the "outside men." The regular men were regularly hired and rehired to work on the docks at the port of St. John's, where one of the requirements to be a "regular man" was to learn the proper drinking behaviour for members of the group.

These patterns were so fundamental to the continuities between nonwork and work roles that in Mars's research the men rarely mentioned the importance of drinking roles without explicit questions because "I didn't think you counted drinking! Everyone always drinks with their buddies!".

The consumption of alcohol is always subject to rules and regulations and breaching those rules arouses strong emotional response. Currently we see such intense emotionality expressed through the Mothers Against Drunk Driving (MADD) campaign. MADD has garnered phenomenal support in American society, to the point of strongly influencing such formal legal controls as blood alcohol levels permissible for drivers and the penalties for exceeding those levels.

Gender issues and alcohol and drug use and abuse have garnered considerable attention recently. Two edited books, for example, address differences between men and women with respect to the proper and improper use of alcohol and alcohol and drugs in different cultural settings.

This genre of research offers much promise regarding its relevance to gender studies generally and, more particularly, regarding an explicit concern that women's drinking and drug use patterns be better documented and understood. Typically the province of sociologists in the past, workplace-based research on alcohol consumption and abuse has been studied by anthropologists at the Prevention Research Centre.

Acculturation and Culture Change Studies

By the 1960s, anthropologists had developed an increased interest in applying their ethnographic and often emic approach in order better to understand the problematic use of alcohol. This was particularly the case with respect to American Indians. As part of the interdisciplinary team at the Tri-Ethnic Research Project at the University of Colorado, Theodore Graves contrasted SpanishAmerican, Anglo-American and Ute Indian groups living in the same area as a means for addressing the question: "Under what conditions is acculturation accompanied by symptoms of social and psychological disorganization and under what conditions it is not?". Stark differences in drinking practices and problems with alcohol as well as other types of social problems existed among these three cultural groups.

Graves and his colleagues combined ethnographic observations with structured and unstructured interviews conducted with a randomly selected sample from the three groups. He found that the relatively unacculturated SpanishAmericans and Indians evidence patterns of alcohol use and abuse distinct from each other. While the Spanish-Americans retained strong controls socially and psychologically, the Indians were weak on this dimension. Furthermore, while the unacculturated Indian groups displayed excessive drinking patterns and problems associated with heavy drinking, the Spanish as a group did not.

With this study, a long tradition in acculturation studies of American Indians and other ethnic groups in the United States was well underway. Paradoxically, acculturation is sometimes hypothesized to be a protective factor against dysfunctional drinking practices and at other times a risk factor. As Graves pointed out in 1967,

Acculturation is obviously not the unqualified evil that some observers regard it. When traditional cultural strategies for personal satisfaction have become inapplicable, a reorientation toward a set of new and potentially attainable goals appears to be a promising path to mental health. Furthermore, where traditional social and personal control systems are weak, acculturation may also serve to promote the development of new controls and thereby make the group better able to prevent disruptive individual behaviour.

Change in drinking patterns under conditions of culture change and/or acculturation has been an ongoing theme of alcohol research in the past three decades. The Institute of Applied Social and Economic Research (ISAER) Alcohol Project in Papua New Guinea is an example of an ambitious two-year research project directed by anthropologist Mac Marshall and undertaken to address a growing problem with alcohol there. A notable feature of the resulting conference and monograph is the fact that most presentations were made by scholars—including many anthropologists—who had conducted indepth ethnographic research at a particular field site on Papua New Guinea. While none of these researchers had intentionally undertaken a study of alcohol and culture, each discovered that alcohol consumption was important enough to collect data on the topic.

In 1982 Marshall edited a monograph on changes in drinking patterns in highland and coastal villages and urban areas of Papua New Guinea and problems associated with drinking. The significance of this project is seen in the fact that it constitutes "the first time anywhere that such a large body of ethnographic information has been assembled specifically in the service of public policy decisions on alcohol".

Drunken Comportment

Two issues—the extent to which alcohol consumption poses serious social problems and the extent to which available alcohol control policies help to stem the tide of those perceived problems—have historical reference in anthropology in the "drunken comportment" concept. In their 1969 book, psychologist Craig MacAndrew and anthropologist Robert Edgerton examine the phenomenon of drunkenness cross-culturally using ethnographic and ethnohistorical data to evaluate several assumptions held about behaviour under the influence of alcohol. First, drawing upon ethnographic data reported from five societies, they evaluate evidence for the "disinhibiting" effects of alcohol. They found that "even during periods of extreme intoxication, the inhibitions that are normally in effect remain in effect. Drunken persons in these societies may stagger, speak thickly and become stuporous, without any corresponding display of changes or-the-worse. In a word, if alcohol were a 'superego solvent' for one group of people due to its toxic action, then the same disinhibiting effect ought to be evident in all people. In point of fact, however, it is not".

In short, MacAndrew and Edgerton call into question the common assumption hold by many people in American society that alcohol serves as an disinhibitor, a point of view often perpetuated by medical science. Thus, even in the midst of increasing cross-disciplinary cooperation, anthropologists continue to find themselves in a position to question seemingly entrenched viewpoints about alcohol and alcohol-related behaviors. Unfortunately, noncritical thinking about alcohol and alcoholism also characterizes some of the research on alcohol. As Heath has succinctly put it, "polemic masquerades as science in much of what is written on alcoholism".

The Disease Concept of Alcoholism

In understanding alcohol abuse—how it develops, how it is diagnosed and how to prevent it or intervene in it—alcohologists tend to rely upon simplistic oppositions such as the nature-nurture or genetic-environment contrasts. Similarly the disease concept is often contrasted with the moral model concept in which alcoholism is seen to stem from weaknesses of individuals, which keep them from controlling their drinking. Historically, American society experienced a long period in which the moral model reigned supreme—during the 150 years of strong temperance and prohibition movements.

With the founding of the Yale Centre for Alcohol Studies after World War II under the leadership of E. M. Jellinek, the disease or biomedical model was developed as a counter to the moral model. The disease model, according to Jellinek, explains alcoholism as a progressive disease with clear symptoms and certain recognizable, inevitable phases. The insistence on fidelity to the disease concept in the treatment of alcoholism, however, is a more convenient, established theory than scientific understanding based upon scholarly work. In reality, American society has superimposed the disease model upon the moral weaknesses model in popular understanding of the etiology and nature of alcoholism. Thus, alcoholics can be held responsible for their addiction based on personality features while at the same time be excused based on a presumed physiological predisposition.

In addition to the confusion around the moral-medical distinction, many wellmeaning professionals, as well as the general public, operate with a vague idea of what is really meant by the disease concept and assume that whatever it is is clearly supported by scientific evidence. Contrary to some popular beliefs, the same symptoms are not always present in all alcoholics or in those seeking treatment for alcoholism. The course of alcoholism varies widely.

Also, from a disease point of view, we might expect that biomedical criteria would be used to diagnose alcoholism. In fact, however, mainly behavioral criteria are applied. Drinking patterns and behaviour under the influence of alcohol figure more prominently in diagnostic criteria established to distinguish moderate, heavy and alcoholic drinking than do strict biomedical indicators such as organ damage and withdrawal signs.

We recently surveyed a group of anthropologists as to their points of view about the disease concept of alcoholism. This is far from a neutral topic—some respondents commented on its sensitivity—but responses were frank. David Strug noted that he had never "felt comfortable with the disease concept of alcoholism based on a biomedical model. The cultural component will always remain a contextual constant that must be considered". In connecting the idea of alcoholism with disease, Merrill Singer concludes that it reflects "broader patterns of medicalization, privatization of suffering and politically-endorsed individualized problem-solving patterns.

While virtually all respondents expressed skepticism about the disease concept, Dwight Heath was the most clearly dubious among these reports: "If alcoholism is a disease, it is a most unusual one inasmuch as an individual can often bring an end to it by modifying his/her behaviour even in the absence of any other intervention. Most of the reasons commonly given for calling it a disease are fallacious".

Interestingly, these opinions are generally consistent with the World Health Organization's position that alcohol problems do not necessarily follow a coherent pattern throughout the general population and do not necessarily have to do with a physiological dependence on alcohol.

Anthropologists insist, as do some other alcohologists, that alcoholism is a complex phenomenon, perhaps having predisposing factors (genetic and physiological) in combination with a series of precipitating factors (including psychol-ogical, social and cultural) contributing to etiology. The need for a biocultural synthesis of studies of individual and cultural variation is compelling.

ANTHROPOLOGICAL WORK ON ALCOHOL

Anthropological work on alcohol and other drugs often

challenges conventional assumptions about substance use and abuse. In this chapter, we delineate this theme by highlighting areas where the uniqueness of anthropological contributions to understanding drug use and abuse is clearly seen. We have not reviewed the entire field of anthropological research in the area; given the enormous body of literature, especially in alcohol studies, such a review would constitute a project beyond the scope of this book. Furthermore, several excellent reviews have been published on specific topics within this research domain. Anthropologists tend to take a different tack in approaching studies of substance use and abuse; consequently, their work is often controversial to policymakers and treatment providers. Controversy—and sometimes skepticism-frequently surround the approach and methods, interpretation of data and recommendations for policy.

This controversy has been recognized within this discipline itself. In certain instances, their orientation is congruent with that of colleagues in other disciplines. In recent years, in fact, some anthropologists have created or joined interdisciplinary research groups and are working as team researchers.

Anthropological interest in studies of drug use and abuse has been apparent for several decades. Until the 1970s, however, most of this research was a by-product of broader ethnographic studies of small societies in Latin America, Africa, Oceania and the North American Indian and Eskimo tribes. In this tradition, data collected on the use of mind-altering substances formed one component of an overall study, such as noting the use of plants for medicinal purposes or describing curing ceremonies where sacred plants were employed.

Cross-cultural research on substances has been conducted through the use of data housed in the Human Relations Area Files, by in-depth analysis of available ethnographic data, through special conferences in which drug use data were presented for a variety of societies, communities, or ethnic groups, or through projects in which the anthropologists conducted a controlled comparison or contrasted different drug use traditions cross-culturally. Thus, it is fair to conclude that as of the early 1970s, anthropology had not yet developed an explicit drug research tradition, especially with respect to abuse of drugs. With increased funding available for such studies and with the expansion of applied anthropology, the situation has changed dramatically over the past two decades.

In response to the AIDS epidemic, anthropologists have become central participants in a major research agenda on the sociocultural context of risk behaviors in an effort to develop effective prevention programs. In fact, anthropologists have been quite successful in obtaining funding for AIDS- and HIV-related projects in large part because of the need to conduct qualitative studies in this area. Increased risk for HIV infection as related to alcohol and drug use has been an important aspect of this research. The field of drug studies has developed into several subspecializations among anthropologists concerned with family, AIDS, treatment and special populations (such as North American Indians, Hispanic groups, black Americans, the working class and women).

Up to the present, anthropological discussions of substance use and abuse are almost always treated separately with respect to alcohol and other drugs. This tendency reflects wider patterns in national and international policy and research, as well as treatment of substance abuse. For example, the National Institute on Alcohol Abuse and Alcoholism was established in 1970 separate from the National Institute on Drug Abuse and the World Health Organization continues to separate alcohol from the "illicit" drugs in most of its working conferences and publications. In anthropology, it is relatively rare for more than one drug to be encompassed in anthropological studies and publications.

Anthropologists in the substance use-abuse field have focused primarily on studies of alcohol, reflecting, in part, the relative order of usage of particular drugs throughout the world: first, ethanol; second, nicotine; third, caffeine; fourth, betel; and fifth, marijuana. Over the past twenty years, however, a substantial tradition has developed in cannabis research, mainly conducted in the Caribbean and Latin

America. Similarly, there has been an increasing interest in tobacco studies and trends in publications and symposia at anthropology meetings seem to indicate that this line of research is gathering momentum. Some research has focused on the role tobacco plays in social relationships. Another theme has been the interaction of transnational tobacco companies seeking new markets in Third World countries. Several researchers with roots in early studies of religious experience and hallucinogens have specialized in the cultural context of the ingestion of hallucinogens such as peyote and mescaline in the past two decades.

Due in part to American society's definition of heroin, opium, marijuana and cocaine use as clearly deviant behaviour and in part to demand for legal and/or clinical intervention, studies on such drugs by anthropologists have attracted considerable attention. The street culture around the use of heroin and the use of methadone in the treatment of heroin addiction has, for example, drawn a kind of interest that is perhaps more curious than serious.

While such anthropological research tended to focus on heroin in the 1970s, it shifted to an interest in crack-cocaine in the 1980s and 1990s, as use patterns and the wider society's concern about the use and abuse of certain substances changed. The typical anthropologist's emic focus on the drug user's cognitive and social worlds is often at odds with the clinician's and policymaker's perception of the' problem. The clinical and policy domains of our society take a more etic stance with respect to use of these "illicit" drugs and thus express impatience, at the very least, with arguments that suggest the use of these substances is not necessarily deviant from the user's perspective. This perspective has characterized much of the anthropological research on alcohol and drugs during the 1990s.

Regardless of the position taken regarding the deviance of such drug use, it is becoming apparent that prevention and treatment programs must do more than remove the drug in order to be lastingly effective. Drug use fits into a cluster of behaviors and beliefs and treatment agendas must deal with that reality in proposing alternative ways of life to a recovering addict. Thus, a lucid understanding of the cognitive and social worlds of drug users is highly pertinent to preventive and intervention efforts.

In addition to twenty years of research on heroin addicts anthropologists have more recently undertaken studies on cocaine, as cocaine has gained popularity as the drug of choice for many Americans. One such study examines the ways in which cocaine users interpret their environment and the resultant influences of the user subculture on drug use patterns. Another combines interest in issues of multiple drug use by investigating cocaine use among methadone clients. Thus far, a small body of research on cocaine exists in anthropology. Because of the relatively widespread indigenous use of kava and betel in Oceania, anthropologists working there have in the course of ethnographic research reported on the use of these substances.

Anthropologists approach research in drug use and abuse by illuminating the cultural context in which they take place. For example, while American society has dramatically moved in an antitobacco direction, some anthropologists have recently called attention to the functional role of tobacco use. Consequently, anthropologists at times have been seen as rabble-rousers and troublemakers who rock the boat of cherished assumptions about the pathology or deviancy of drug use. For example, anthropologists have been taken to task for overstressing the functional role of alcohol in culture and ignoring its dysfunctional use.

Three decades ago, David Mandelbaum clearly articulated the anthropological slant on functional and dysfunctional use of alcohol: "Drunkenness cannot be understood apart from drinking in general and drinking cannot be understood apart from the characteristic features of social relations of which it is part and which are reflected and expressed in the act of drinking". More recently, Mac Marshall has similarly stressed the essential value in examining deviant drug use within the context of normal patterns: "All the ethnographic accounts show the necessity of understanding the variety of normal drinking styles in any social setting before attempting to deal with abnormal (or addictive) drinking".

As colleagues in other disciplines have taken a greater interest in the role of culture in drug use and abuse, they have understood and applied the concept of culture in ways that are sometimes discrepant with the concept in anthropology. Specifically, in the minds of many scholars outside anthropology, "culture" has become synonymous with presumed membership in a particular ethnic group, nationality, racial group, religious affiliation, class and so forth as related to particular drug use and abuse patterns.

This perspective stands in contrast to the more traditional anthropological view of culture as a dynamic process through which individuals and societies learn the sum total of their society's behaviors and associated belief systems, including those encompassing drug use practices and beliefs.

MIND-ALTERING PLANTS IN SOCIETIES

Fieldwork focusing on the magicoreligious use of hallucinogenic plants is perhaps the (nonalcohol) drug field's best representative of traditional academic anthropology, as distinct from the growing applied area of the discipline. In the case of the sacred plant hallucinogens—such as peyote, mescaline and mushrooms—traditional ethnographic research is not typically undertaken to improve treatment or to inform policy. Its aim is to broaden the understanding of the human experience by illuminating the perspectives and experience of others.

The vantage point of cultural relativism has especially influenced this line of anthropological research. While this influence has helped provide fresh insight into the cross-cultural use of hallucinogens, it has also reinforced the controversial posture of much anthropological research on drug use.

Cross-cultural assessments of the role of hallucinogen use illustrate how controversy may take form. While most people in Western society typically believe that the use of hallucinogenic drugs is dysfunctional, ethnographies of tribal cultures have often shown that these substances have been used for generations without disruptive effects to the society. In Western and some non-Western societies, drug users have divergent experiences apparently related to wider cultural differences.

Comments from Marlene Dobkin de Rios illustrate this point: "Lacking specific cultural traditions of drug use which programme their experience, Westerners often report idiosyncratic patterns. There seems to be good evidence that in a society where plant hallucinogens are used, each individual builds up a certain expectation of drug use which, in fact, permits the evocation of particular types of visions".

Other ethnographers have written similar accounts of the user experience in traditional societies, noting the power of culture to shape the drug encounter, to "determine the nature and intensity of the ecstatic experience and how that experience is interpreted and assimilated".

Both Dobkin de Rios and Peter Furst find such elements as ritual preparation for and ritual control of the drug experience common to societies with magicoreligious traditions of hallucinogen use. In these societies, the substances are used to evoke visions or stir insight; as sacred plants, they rate a level of regard not accorded recreational drugs in complex societies.

In comparing mainstream American and North American Indian societies, Weston La Barre observes that in some tribal groups, American Indian adolescents receive drugs and guidance in their use from adults through socially established and respectable channels.

This procedure differs considerably in intent from that of many other American adolescents who are resisting the influences of the dominant culture. In the former case, the adult order is reinforced; in the latter, it is assailed. La Barre draws on a body of anthropological knowledge about hallucinogens to connect "altered states of consciousness," including dreams, hallucination and similar states with the origins of religion. He finds such altered states to be genuine human universals and asserts that the experience of shamans while in such states, often induced by hallucinogens, shapes the revelations on which religions are based.

THE "FIREWATER MYTH"

The "firewater myth" is a major arena where the genetic versus environment debate with respect to population-level differences has been waged for decades and continues to attract the attention of anthropologists and others concerned with these issues.

Here conventional wisdom states categorically that North American Indians cannot handle alcohol because of their physical constitution, an idea going back to the nineteenth century. Joy Leland has traced the history of the argument and examined the available published data as of the mid-1970s.

One explanation for widespread disagreement as to the extent of alcohol addiction among American Indians is the lack of a set of agreed-upon symptoms of alcohol abuse. Leland points out that we are far from having such a list of symptoms for the "dominant society," let alone distinctive cultural groups such as American Indians.

In her review, Leland tabulates the presence or absence of forty-four symptoms of alcohol addiction proposed by E. M. Jellinek that are documented in the available literature on North American Indian groups.

Using the symptoms checklist, it was not possible to confirm the firewater hypothesis. On the other hand, the reverse of that hypothesis cannot be decisively supported or discredited. Much more scrutiny is necessary before any firm conclusions can be drawn about the extent of alcohol addiction among American Indian groups, let alone the potential biocultural basis for such addiction.

The term firewater myth uses the word myth in the sense meaning "misconception," or a notion based on something other than fact. Myths of this kind can mold attitudes and thus actions, as can myths of a more traditional stripe, such as creation myths: "Myths are powerful influences in human affairs: they condition situations, their preconceptions create consequences".

EMERGING DIRECTIONS FOR THE FUTURE

In addition to maintaining its well-established course of explicating the cultural context of drug use and abuse, anthropology might target three areas for special attention: culturally focused treatment and research, biocultural studies of alcoholism and broadening anthropology's impact on the alcohol and drug field.

TREATMENT AND RESEARCH

Anthropologists have argued that knowledge about cultural variation in drug use and abuse patterns—as well as the wider cultural context—is important for planning effective prevention, intervention and treatment strategies. In fact, Joan Weibel-Orlando asserts that "we accept, as a disciplinary mission, the role of revealer/advocate. We advocate the right of a people to heal themselves in any manner they see fit. To do so is a noble cause consistent with longestablished anthropological ethics and belief in the cultural relativity of all institutions including those that attempt to heal and bring people back into 'balance'".

What is the evidence for the efficacy of treatment programs designed with the cultural background of the clients in mind? H. K. Heggenhougen points out that traditional systems of healing have been applied to addiction for some time. In Hong Kong, for example, acupuncture has been used to treat withdrawal. He also notes that conventional treatment programs do not have outstanding success in treating addiction and that they are not available to all who need them. Although the efficacy of alternative treatment for addiction is still uncertain, he argues that treatment provided through traditional or indigenous medical system is a needed resource.

Joseph Westermeyer has made a similar point. He studied Laotian opium addicts in voluntary treatment at a Buddhist monastery in Thailand. Alternatively, treatment at a medical centre was available, where patients were slowly withdrawn from opium through methadone therapy. Clients of the monastery's programme underwent "cold turkey" withdrawal, yet many addicts continued to seek it out because of its "traditional and reassuring location". Some difficulties arose at the monastery due to language barriers between the

Thai monks and some of the Laotian tribal people: clients at the medical facility complained of the same problems. Both groups, however (monastery and medical centre), had good overall feelings about their treatment experiences. Both have similar long-term outcomes in client abstention. The monastery had a substantial cost advantage over the medical facility but lost favour because of the mortality among older addicts in the withdrawal. A combination of the best features of both might be optimal.

Many anthropologists have posited the increased efficacy of treatment programs that take into account the cultural milieu of patients. This hypothesis, however, has not been solidly tested. Weibel-Orlando—after acknowledging her own strong support of this position, especially with respect to treatment programs for North American Indian groups—then calls into question the research base for the argument: "Most of our enthusiasm for indigenous curing strategies as viable contemporary alcohol and drug interventions is based on anecdotal materials. More systematic and observational investigations of the efficacy of such interventions are needed".

Biocultural Synthesis of Alcoholism Etiology

Discussion of the firewater myth debate leads directly to the question of why anthropologists have not led the way in studies combining biological and cultural factors in presumed population-level differences in the incidence and prevalence of alcoholism. On an individual and family level, a biocultural synthesis might help unravel some of the thorny questions around the interplay between predisposing (genetic) and precipitating (personal, social and cultural) factors in alcoholism etiology.

With its integrative approach, anthropology would seem to be in a particularly good position to design and conduct such studies. James Schaefer proposes two types of studies that would address these issues. He points out the need for both general population and large family pedigree studies. If both were to be conducted in a well-conceptualized manner, together they could provide fresh insights into the "biophysiological or sociopsychological predispositions to alcoholism".

With respect to existing data about differences between North American Indian groups and other ethnic groups, Schaefer strongly contends that "we are a long way from having conclusive evidence of differences in ethanol metabolism by racial group". In addition to the methodological problems with existing studies, he highlights the basic point—too often overlooked in these discussions—that hypothesized physiological differences between groups in ethanol metabolism cannot be automatically presumed to place the group at increased or decreased risk for alcoholism; certain questions must first be addressed. For example, what are the consequences of relatively greater physical discomfort after drinking and what are the likely consequences of more rapid metabolism of ethanol upon actual drinking patterns for a group? Do these differences provide increased or decreased protection from possible addiction?

"A paradox thus emerges. Biophysiologically based hypersensitivity reactions may be 'protective'; however, in some cases social, psychological and cultural factors may transcend the 'protection.' Indeed, a wide spectrum of behavioral evidence points to increased amounts of cultural stress, powerlessness and anxiety as conditions which exacerbate alcohol use". The family pedigree study approach would be a fruitful direction in attempting a biocultural investigation of alcoholism etiology. To date such studies have focused on biophysiological measures, drug use histories and social-demographic variables. The cultural orientation of anthropology should be a core component of such studies.

Increasing Anthropology's Impact

Much anthropological evidence has not been incorporated into professional and lay understanding of alcohol and drug use and abuse. A major gap exists between research findings and the application of results. Why might that be the case?

One, the issues are complex and it is truly difficult to sustain the attention of a reader or listener long enough so that he or she can make an informed decision. The more that anthropologists—and other scientists—clearly present the evidence for and against a position and the more that it can be interpreted in real-life terms, the better the chances are to make changes in public opinion. Second, conclusions from anthropology often seem to go against the grain of common sense, making the business of influencing people's thinking an uphill struggle. An open acknowledgement of those differences is perhaps more constructive than assuming audiences will be swayed by a solid argument.

Third, anthropologists may not be placing enough importance on presenting their arguments to the right audiences. It is essential that anthropologists not limit their written and oral presentations to other researchers. Courses and workshops in the area of alcohol and drug studies can, for example, draw students who are hungering to explore the relationship between culture and biology in the development of drug-related problems.

They also provide a good forum for expanding the dialogue between researchers and clinicians. Publication in popular contexts and delivering public talks are also critical to advancing the general understanding of alcohol and drug use and abuse. Anthropology can and should be an active force in expanding informed discourse on alcohol and drugs, leaving its distinctive imprint on the process of inquiry.

Chapter 7

Medical Anthropology

Medical anthropology is primarily an applied subdiscipline, as should be apparent from the materials covered in this book. The roots of the subdiscipline reach back to an intellectual, academic interest in describing and understanding the ways in which various non-Western peoples have explained illness and given treatment to the sick; but the preponderance of research in the 1980s and 1990s has centered on pragmatic issues of improving the health and health care situations of contemporary people, both "Western" or "non-Western." Health problems throughout the world constitute a sector of applied research that is by nature interdisciplinary; most health issues require data from the biological sciences, clinical medical practice and the social-behavioral sciences.

Research in health problems often involves other types of expertise as well; for example, the role of entomology is very important to understanding various vector-borne diseases such as malaria, typhoid and dengue fevers and the growing interest in research on health care systems requires information from economics and political science. Although there are many instances of research in which individual anthropologists, medical doctors, or biologists "did it on their own," such solo performances are increasingly suspect, given the complex data involved in health issues.

The interdisciplinary nature of the illness and health care sector is partly responsible for the fact that methodological issues are strongly affected by national and international agencies and other organizations that sponsor research. In the

United States a very large share of health-related research is funded by the National Institute for Mental Health, National Institute for Drug and Alcohol, National Cancer Institute, National Institute on Aging and other federal agencies.

On the international health scene, the World Health Organization (WHO), the U.N. International Children's Emergency Fund (UNICEF), the U.S. Agency for International Development (USAID) and a variety of other organizations sponsor health-related research. Proposals for research in any of these national and international agencies are judged by interdisciplinary review panels, often (but not always) dominated by biomedical scientists.

These factors have had considerable influence in shaping the directions of research methodology in medical anthropology. Also, increasing numbers of medical anthropologists are based in medical schools, schools of public health and other health agencies, in which collegial relations are strongly interdisciplinary. On the other hand, a substantial portion of research in medical anthropology continues to be funded by the anthropology division in the National Science Foundation, the Wenner-Gren Foundation and other sources in which the review panels are primarily anthropologists.

These anthropology-oriented sources are especially likely to be tapped for funding by medical anthropologists whose primary affiliations are in anthropology departments. In such cases the research designs and other methodological features are somewhat less affected by the interdisciplinary (particularly the biomedical) realm of discourse. It is probably fair to suggest that such "anthropology-oriented" medical anthropology is less often applied in nature. However, one can find many exceptions to these patterns.

The growth of medical anthropology over the past two decades has been especially evident in the applied, interdisciplinary realm. In applied research, the solutions to specific practical questions about health and illness are the central concern and development of theory plays a secondary role. Theoretical concerns are not totally ignored, but the areas of theoretical interest are often in "theories of the middle range," where conceptual issues are strongly intermingled with methodological strategies. Medical anthropologists often pay lipservice to aspects of grand theory, but the research is usually at a considerable remove from broader theoretical abstractions.

In any case it is possible to examine a great many issues in the methodology of medical anthropology without direct commitment to a particular theoretical position. In fact, much of anthropological method is essentially theory-less, in the sense that the basic methods of data gathering are the same regardless of the theoretical system adopted by the investigator. In field research it appears that practically all anthropologists use a mixture of interviewing (both structured and unstructured) plus direct observation (again, both structured and unstructured).

Specific questions asked and specific targets for observation differ, depending on theoretical interests, but the processes of data gathering are broadly similar regardless of theoretical orientation. It is in the language of theoretical discourse that anthropologists differ markedly, even when discussing basically similar data.

This is not to say that two different theoretical discourses necessarily disagree with one another; quite often the different theoretical vocabularies are in some sort of complementary, noncontrastive relationship. Our examination of field methodologies in medical anthropology will be presented in a generally nontheoretical, or theory-neutral, manner. However, certain methodological tools and techniques will be presented with reference to particular research examples, which may include some of the theoretical language of the authors of the research.

BASIC RESEARCH IN MEDICAL ANTHROPOLOGY

Research design becomes specific when we address specific questions. In much of the research in health care, as carried out by medical anthropologists and others, the basic questions very often consist of variations on three main (applied) thematic areas:

- Descriptive questions. What do people believe about

illnesses—their causes and treatments? What do they do (e.g., behaviors that increase or decrease risks of illness; specific treatment-seeking behaviors)? What are the characteristics of the health services and systems in which these actions occur?

- Analytic questions. What factors and systems explain variations in beliefs, actions and outcomes?
- Intervention-oriented questions. What are the ways to change and improve the health of particular populations, in terms of system changes, changes in knowledge and actions and prevention of illness-causing conditions?

These are not the only types of basic questions in medical anthropology, but a very large share of research is focused on specific issues related to these fundamental concerns. In a great many instances of research, then, the dependent variable of interest centers on a particular illness or condition--often the actual frequency of the illness. A great deal of medical and health care research, after all, is directed to lessening the frequency (incidence or prevalence) of specific illnesses.

Just as frequently, however, the dependent variables centre on people's choices of forms of treatment. Who uses "indigenous" treatments versus "cosmopolitan" resources to "do something" about a particular health problem?The independent variables are much more varied and they are by nature more directly reflective of basic theoretical approaches. The following hypotheses concerning "causes" or "factors" affecting treatment choices are all in the same grammatical form and can be examined with basically similar methodology, but they reflect different theoretical assumptions and language:

- People [in community x] avoid cosmopolitan health care because of their traditional health beliefs.
- People [in community x] choose indigenous versus cosmopolitan health care depending on their assessment of the severity of the illness and their ability to meet the costs of the specific health care.
- People [in community x] will go to cosmopolitan

health providers and will follow the medical advice to the extent that the information fits with their explanatory models of a specific illness.

- People [in community x] see health care as a political expression and they choose or reject cosmopolitan health cam on political and ideological grounds.
- People [in community x] are likely to be more accepting of the newly introduced primary health care (cosmopolitan) if they have the opportunity to participate actively in the planning of the health service system.

Although these are only a small fragment from all possible research statements, generalizations, or hypotheses, they are useful in illustrating ways in which different researchers, with different theoretical approaches, often have the same implicit or explicit dependent variable (a behavioral outcome) in mind and they will use basically similar methodological approaches to gather the relevant data.

In the five hypothetical cases, each researcher would presumably collect data on people's choices of health care alternatives, through direct observation or interviewing and would also collect information about the network of independent variables specified in their particular theoretical model.

Some researchers may adopt a strategy of direct observation plus unstructured interviews; others might rely mainly on quite structured interviews; still others will opt for various mixtures of quantified and qualitative data gathering.

CONCEPTS AND DEFINITIONS

Before exploring the wide-ranging inventory of research designs in medical anthropology it will be useful to present some basic definitions of terms that are central to methodological discussions. These terms play a central role in the structure of research proposals, so they constitute a key element in the vocabulary of "grantsmanship," as well as in the analysis of different approaches to theory building and problem solving in medical anthropology.

Data

Data are the recorded results of empirical observations in fieldwork, both quantitative and qualitative. All field notes are data; the recorded responses on structured interviews and their transformations into computerized data sets, are data. Photographs, documents and other physical materials also constitute data. Note that we use the term data to refer to both the physical materials (including tape recordings) and the variables or "themes" or other attributes extracted from the primary materials. Sometimes we use the term raw data to refer to the actual physical materials, including unprocessed field notes.

Variables

A dependent variable is an outcome or condition or phenomenon that is to be explained or accounted for or predicted, by, reference to presumed "causal factors," "prior conditions," "determinants," "disposing features," or other conceptualizable antecedents.

An independent variable is any presumed "causal factor," "prior condition," "determinant," "disposing feature," or other conceptualizable antecedent that is thought to account for, predict, explain, or contribute to the existence or specific form of an outcome or condition or phenomenon.

In experimental and quasi-experimental research designs, it is almost always the independent variable that is manipulated. If a research project has an experimental and a control group, the nature of those two groups constitutes, or embodies, the major independent variable. (Although many researchers have come to associate the notion of variables with statistical analysis, all empirical research can be usefully conceptualized in terms of variables, however implicit they may be in the actual research reports. Thus, data concerning particular variables may be "highly quantitative" or quite qualitative in presentation.

Hypotheses

A hypothesis is a more or less explicit statement of a

hunch, expectation, or prediction of relationships or patterns that one seeks to test or examine in the course of a specific research project. Hypotheses, like operationalized definitions, are best seen as aspects of specific research projects.

Methodology

This concept refers to the logic-in-use in any research project whereby "raw" empirical observations are assembled and transformed into successively more abstract descriptive and analytic statements. Methodology may be thought of as a series of transformational rules and processes (including definitions of key concepts) that guide data gathering and relate the resulting data systematically to the hypotheses and other conceptual models in terms of which research results are expressed. Statistical procedures are one type of transformational system for arranging complex arrays of numerical data into patterns that can be expressed as theoretical models.

Models

A model is any representation of the interrelationships among a series of variables or constructs in a research domain. A model is thus an analogical, simplified, physical representation of the phenomenon in a particular instance of research. Commonly encountered models include maps, diagrams, scale models of physical things, as well as verbal descriptions that aptly portray essential elements of a complex domain.

A famous model is the physical representation of the double helix used by the biologists Watson and Crick in arriving at the description of the DNA molecule. In anthropology, particularly in earlier decades, the most commonly encountered models were representations of kinship terminologies. For our purposes, the term model is the meeting ground between the theoretical and methodological realms of discourse.

A model embodies the elements derived from a particular theoretical perspective. Thus, the terms or features of a model are simplified portions of a general theory. At the same time, the model includes the elements or details about which specific data are to be gathered in a research project. Each element or concept in a model requires some sort of "operationalized" representation in the research activity.

Operational Definition

An operational definition (of a variable) is a statement of specific datagathering procedures that produce indicators for a given independent or dependent variable. The procedures often include statements of cut-off points, such as, "High blood pressure will be defined as a measured systolic pressure above 140 and/or diastolic pressure above 90." Here is another example: "Socioeconomic status in this research was dichotomized into two groups, landowners (having more than I acre of arable lands) and the landless."

Some researchers appear to consider the idea of operational definitions as referring only to quantitative research. However, the logic of this concept is the same, whether quantified or not. All concepts reported by researchers arise from data of some sort. The reader of any research can always ask, "What data serve as evidence for this particular statement?" Much of the writing in anthropology, including medical anthropology, presents information without specifying details of research methodology. Often we are left to guess at the operational definitions. But they are still part of the research structure, even if they remain unreported.

Triangulation

In this strategy in ethnographic research, data concerning a particular topic are gathered from more than one source, or using more than one technique, so that systematic comparisons (and possible corrections) can be made. Examples of triangulation include the systematic comparisons of the statements made by different key informants and the comparison of key informant statements with the results of structured quantitative surveys. Another common form of triangulation that has come into vogue is to compare focus group discussions with key informant interviews and/or quantitative survey results.

BASIC UNIT OF ANALYSIS

In most situations, the people of interest to medical anthropologists experience their health and illness and make decisions about health care in the context of the household or coresidential group. The specific operational definition of household may vary for different populations, but the general term refers to a group of people living together in a single domicile, sharing food and other resources, whether or not consanguineally related.

Often researchers seek to delineate households as the people who eat from the same pot, even in cases in which more than one such cooking-eating group may be found within a compound or other complex domicile.In most community-based studies, the common practice is to carry out some sort of census or enumeration of all the households, in order to define the universe (the population) from which samples may be selected. Even when research is mainly participant observation and unstructured interviewing, it is good practice to establish a baseline census. When large numbers of households are involved, the basic census is limited to a small list of key questions:

- Name, age and sex of each person (and their relationship to household heads).
- Ethnic identifications of household heads.
- Occupations of adult members (including cash crops).
- Education of adult members.
- Religious affiliation of adult members.
- Physical indicators of house quality (usually number of rooms, floor material, roof and walls, number of windows).

The physical indicators of house quality are useful as an approximate measure of socioeconomic status.

In addition to these items, each household (and usually each individual) should be designated with a unique identification number, to relate all subsequently collected information and the selection of research samples, to the correct units. Commonly the identification number is composed of community, household, individual, as follows: 01(community)/ 001(household)/ 01(individual) = 0100101 (the first person in the first household in the first community).

In many countries the health ministry or one of the government health research institutes may have a standard census form that it wants all researchers to use. Such "nationwide" formats have the advantage that they permit some comparisons of the specific research population with other areas of the country. On the other hand, the standard forms often include portions that are obsolete or inappropriate for given regions. If possible, researchers will use the official protocol, with additions and modifications to fit local conditions.

If resources are available for gathering more information in each household, the additional items will reflect the specific research concerns, as well as special ecological and other local features important to specific health-illness issues-for example:

- Sources of water supplies.
- Sources and types of fuel and cooking facilities.
- Latrine, toilet facilities.
- Immunization status of children and women.
- Physiological status of women (pregnant, etc.).
- Usual source(s) of health services.
- Labour migration status of family members.
- Recency of arrival to this area and community.
- Community of origin of adult members.
- Foods produced by household.
- Animals owned or maintained by household members.
- Ownership of selected consumer items (radio, television, vehicles, etc.).

Many other items can be added to the list of basic questions concerning the universe of households. However, very few researchers can afford to collect even this much information from all households in their study communities. Quite often a researcher (or research team) will direct the extra questions to a subsample—perhaps every tenth household of the overall census. In this way at least approximate frequencies can be obtained for a variety of features that can then be studied in greater depth as research progresses.

The census, or enumeration, of all households in a study population has other functions besides the collection of specific data. Regardless of whether the process occurs at the outset of research or later, the census is an important opportunity to introduce the research group and purposes of the research, to all households in the area.

In addition to the information about the project, each household can also be given information about any expected health interventions connected with the study. The census enumerators can distribute health education leaflets and information about clinic times and places and can recruit volunteers for local health committees. Census contacts can often help in identifying potential key informants, such as local healers.

THE ANTHROPOLOGICAL APPROACH

Compared to most other disciplines, the hallmark of anthropology, medical anthropology included, is the so-called holistic approach. This takes many forms, but in most research there is the assumption that for any particular outcome or phenomenon to be explained, there are a great many interrelated factors at work. In practice, this means that medical anthropologists are likely to collect a great deal of data about economic features, social relationships, cultural belief systems, political processes and other aspects of a community, even if the research intention is focused on a specific health question.

Ibis holistic perspective often leads anthropologists to be highly critical of other disciplines when they appear to adopt single-factor explanations or seemingly simple explanations for illness conditions, health care responses and other issues.

The holistic perspective has important effects on research design. Whenever numerical analysis is involved, medical anthropologists are likely to be concerned with a large number of variables, requiring fairly complex statistical procedures. Also, attention to large numbers of factors, or variables, requires a considerable investment of time for each case, patient, illness episode, or other unit of analysis. The time limitations (and limitations of personnel) in turn constrain the anthropologist to limit sample sizes severely. The typical project in medical anthropology is likely to have much smaller samples than, for example, corresponding research projects by epidemiologists, sociologists and demographers.

Another hallmark of medical anthropology is the central role played by the concept of culture. In recent years many other types of researchers have come to recognize the importance of cultural differences and cultural effects in relation to health issues, but for anthropologists, the concept has much greater importance in shaping the directions of research.

Earlier, before the subdiscipline of medical anthropology came into being, many anthropologists who studied matters of health and illness among non Western peoples regarded the detailed description of traditional healers and cultural beliefs about illness to be the primary ethnographic objective. They often paid little or no attention to instances in which people used cosmopolitan medicines and practitioners. That is, the primary emphasis of earlier work was on the traditional belief system rather than on actual behaviour. In such studies, then, "the culture" was seen as the sole topic of data gathering.

The concept of culture has now assumed a more modest place in the theoretical and methodological works of many medical anthropologists. "Culture," and cultural differences, have come to be seen as one major cluster of variables, along with complex networks of other factors that account for, or explain, actual behaviors. Ibis shift in the use of the culture concept constitutes a major achievement in anthropological methodology and metatheory.

The development of the idea of culture as distinct from behaviour has made it methodologically possible to speak of (and carry out research on) the variable effects of culture on behaviors. Not all anthropologists share this definition of culture, but there is a widespread tendency to consider culture as idea systems, systems of symbolic meaning, or other variations in language that all focus on people's mental processes. For example, a widely cited book by Arthur Kleinman states that "we can view medicine as a cultural system, a system of symbolic meanings anchored in particular arrangements of social institutions and patterns of interpersonal interactions".

Similarly, Horacio Fabrega, in his book Diseaseand Social Behaviour and Social Behaviour, commented that "illness, for example, offers an additional opportunity to study how behaviour is structured and organized by underlying cultural rules". He then noted that "culture by definition represents a 'man-made,' socially relevant, experientially derived set of rules for living."

Regardless of researchers' specific definitions of culture, one of the central contributions of anthropology to applied studies of health issues is the delineation of the complex ways in which cultural belief systems interact with other factors in affecting rates of disease, definitions of illness, differential responses of illness and other outcomes of interest. Although other disciplines pay some lip-service to the idea of culture in relation to health and illness, medical anthropologists are thought to be the methodological experts in the study of cultural factors.

To a considerable extent, the continued increases in acceptance of medical anthropologists in the interdisciplinary community of health research are due to increased recognition of the cultural factor as crucial to understanding all aspects of illness and health care.

The concept of culture has led to a generally accepted distinction between disease and illness. Illness refers to the culturally defined feelings and perceptions of physical and mental ailments and disability in the minds of people in specific communities. Disease is the formally taught definition of physical and mental pathology from the point of view of the medical profession. Both terms are, of course, "culture." The terms refer methodologically to the contrasts between two distinct cultures that meet when patients interact with physicians, whether in modern urban settings or Third World health systems.

A large share of the research in medical anthropology of the 1980s and the 1990s has focused on situations of cultural pluralism, in which populations with various indigenous health cultures are in more or less extensive contact with the trappings of cosmopolitan health culture. Accordingly, their cultural systems (or "rules for living") include beliefs and rules about the introduced cosmopolitan medications and practitioners, intermingled with the cultural ideas concerning the indigenous healers and treatments. Studies of health care and health issues in urban communities in North America are set in a context of cultural pluralism—as most "mainline" and middle-class people are aware of various alternative health care choices.

CULTURAL VIEWS OF ILLNESS

Arthur Kleinman's formulation of explanatory models (EM) of illness has taken a central place in research on specific sicknesses, as medical anthropologists and others have sought to present a coherent picture of the specific cultural features that affect peoples' health behaviors. The explanatory model for a particular illness consists of:

- Signs and symptoms by which the illness is recognized;
- Presumed causes of the illness;
- Recommended therapies;
- The pathophysiology of the illness;
- Prognosis

As Kleinman points out, individuals are likely to have quite vague and indefinite models of explanation for their illnesses, depending on past experiences of the patient and her or his circle of kin and friends. On the other hand, some individuals in any given community have quite coherent explanations and expectations concerning a specific illness; and the "experts," the healers in the community, would probably on average have more coherent definitions than laypeople, with regard to illnesses and the relevant therapies. In any case, recent research by medical anthropologists has frequently made use of the EM construct as a focus around which a variety of questions can be raised concerning treatment behaviors and other features. Some of the examples of research designs described below focus on methods for systematic relating of explanatory models and treatment-seeking behaviors.

In the past two decades researchers have increasingly recognized the methodological importance of intracultural and intracommunity diversity in people's beliefs and practices. Ibis tendency in research arose in part in relation to the growth of cultural pluralism, especially in matters of health and illness. Medical anthropologists have come to realise that even in seemingly "isolated" communities, individuals and families differ in their degree of adherence to traditional, indigenous health practices, as well as in their attitudes about medical-health ideas and materials newly introduced into their regions.

As a direct consequence, researchers have recognized the need for representative samples of individuals and households, from whom cultural data are collected. The older ethnographic methodology, based on a few selected key informants plus participant observation, is not entirely abandoned, however. Indepth interviewing of key informants, along with participant observation, are still essential aspects of anthropological research, particularly in early, exploratory phases of study.

The qualitative, descriptive materials from this ethnographic work are essential for making sense of the more quantified materials gathered from samples of observations or structured interviews.

RESEARCH DESIGN

At the outset of research on health issues a major decision must be made: to focus the study on cases and events in clinical (health service) settings or to define the research population as community based. That decision has major implications for both qualitative and quantitative aspects of research design.

Some researchers have found it useful to combine clinical and community-based samples. One very useful model is to start with a community in which cases of illness are identified. Differences between the users of health services and the nonusers can be explored in detail. In addition, the interactions in health care settings can be studied in the user subpopulation.

A clinical population can be defined as any group of patients, clients, or cases selected from the persons found at a particular health centre, hospital, or individual healer's location. Clinical populations are selected for research whenever a portion of the research issues focus directly on the activities of the clinic or when it appears that a substantial part of the "cases" of a particular illness are to be found at the clinical setting.

Medical anthropologists have focused increasing attention on the cultural systems, technical workings and other aspects of health care in hospitals and other health care settings. Direct observation of practitioner-patient interactions has become an especially important methodological focus, as researchers seek to define more precisely what really happens in therapeutic encounters. K. Finkler spent two years observing physician-patient interactions in a large hospital in Mexico City. She was present at 800 consultations and she collected detailed narratives from patients about their illness and systematic data concerning their family backgrounds. Follow-up visits were made to the homes of 205 of the patients to get fuller documentation of their illness experiences and perceptions of their interactions with the physicians.

Her massive data collection also included in-depth interviews of seventeen physicians concerning their treatment philosophies and practices. Hospital records provided further depth of information. Earlier, Finkler had carried out systematic observations among healers at a spiritualist temple in a rural region in Mexico. Her data in that study included 1,212 healer-patient interactions. Because of the similarities of data collection methods in the two studies, she was able to make systematic comparisons between the systems of treatment, identifying broad similarities and significant differences.

She noted that "both Spiritualist healers and physicians impose a mind-body dualism on their patients." Also, "In both regimens, the patient takes the role of a passive recipient of the practitioner's ministrations and in both regimens, the practitioners require their patients' compliance reprimanding patients for not having followed prescribed treatments".

On the other hand, Finkler found major differences between the two healing systems: the explanations of illness causation (and diagnoses) are very different; recruitment to the healing role differed greatly; treatment repertoires of spiritualist healers were usually more complex than those of the physicians, who relied mainly on medications.

Contrary to widespread belief among anthropologists and the general public, the physicians she observed spent almost twice as much time with first-time patients than did the healers. On the other hand, "Perhaps the most crucial difference... is this: [spiritualist] healers resolve conflicts for patients that physicians cannot because the biomedical script requires physicians to focus on discrete physical pains".

Studies of provider-client interactions have been directed to the work of other practitioners besides doctors and healers. Rayna Rapp conducted an extensive study of genetic counselors, during which she "observed five genetic counselors working for New York City's Department of Health during their counseling sessions... sitting in on more than 200 intake interviews".

Where specific aspects of the client-provider interaction are studied, the sample unit is often the specific encounter rather than the population of individuals. Accordingly, in some cases the sampling frame is specified as "all clientprovider interactions occurring during period," and a system of randomizing can be applied to the time periods themselves.

A study by Trevathan of childbirth events in a bicultural community provides another illustration of research where the data can be gathered only in a clinic setting. In the case of childbirth, the significant questions often centre on the expectations of mothers in relation to a particular clinic regimen. Accordingly, Trevathan selected a birth centre with a large flow of clients. She enrolled in the one-year midwifery training programme of the birth centre, after which a study of mother-infant interaction was initiated. "Every woman who registered for prenatal care at the Birth Centre and whose delivery was expected between October 1978 and May 1979 was informed of the 'bonding study.' Volunteers were also recruited during childbirth education classes. In the eight-month period, 152 women agreed to be in the study, approximately 50 percent of all those who delivered during that time period".

In this example, focus on cases in a particular clinical setting, where the researcher was a participant, permitted her to maintain close control of the research environment. On the other hand, the generalizations (e.g., concerning differences between Spanish-speaking and Anglo mother-infant pairs) cannot be extrapolated to the general population.

Studies based on clinical samples are often limited by the number of patients in a particular facility. For example, Cohen and colleagues compared and contrasted the explanatory models of diabetes patients with those of clinical staff in a diabetes clinic of a large midwestern university hospital. Their samples consisted of thirty-nine diabetes patients and fifteen professional staff.

Despite the small sample sizes, the study shows interesting areas of discrepancy between the diabetes patients and the clinical staff, particularly in their interpretations of etiology, severity and pathophysiology of the illness. Patients and clinical staff were in close agreement concerning the appropriate treatment for diabetes. Studies focused on cultural patterns such as explanatory models of particular illnesses can often accomplish their objectives using rather small samples.

In the cases mentioned, the clinic populations were appropriately selected because of the nature of the research topic. However, clinic populations should never be considered as representative of the general (community-based) population. In almost every case, a particular hospital or other health setting receives only a selected, nonrandom portion of the population that exhibits a given illness or condition. Other cases may remain home, untreated; still others are found at the various alternative treatment facilities. Even an exhaustive tally of all cases in all facilities does not produce a representative picture of a given health problem, except perhaps with illnesses so severe and so clearly identified, that nearly all of them can be found.

Clinic-based samples, if used as the sole data collection strategy, also have another potential weakness. Patients at health facilities appear as individuals, separated from the family networks in which they normally reside. Full, holistic understanding of people's expectations and reactions concerning illness and health care requires consideration of the household setting as it affects cultural responses. Thus, researchers such as Finkler have often carried out follow-up interviewing of patients in their homes, after observing their interactions in the clinic setting.

Generalizations about the frequencies of health care problems and choices of treatment require sampling from the relevant community population. Epidemiologists often refer to that community-based population as the "denominator," which is essential to study if one is interested in precise estimation of rates (or changes of rates) of particular illnesses or health care practices.

CLINICAL SAMPLES WITH MATCHED CONTROLS

Generalizations from clinic-based samples can often be greatly strengthened by selecting a control group from the same population that the clinic patients represent. K. Finkler, in the study of spiritualist healing mentioned above, introduced this method in her study of the patients. The data from the clinic (temple) sample were systematically compared with a control group (N = 372) "geographically matched with subjects interviewed in the temple corresponding to the villages [from which the patients originated]". Use of the control group permitted Finkler to state that the regular clientele of the temples did not differ significantly from the general population in perceived illness.

STRATEGIES FOR COMMUNITY-BASED SAMPLES

Most research in medical anthropology has been structured in terms of communities or (sometimes) communities within communities. One or more communities in a particular region are chosen as primary sites for research, usually (not always) because of the prevalence of a particular health issue or problem in the selected region. Once the community or communities have been selected, sampling and other aspects of research design depend a great deal on two main factors:

- The nature of the specific health problem addressed
- The geographic characteristics of the communities.

Community-based research is particularly congenial to medical anthropologists because the holistic methodological perspective requires a research context in which the field researcher enters into fairly long-term contacts with the people and is able to combine a great deal of firsthand participant observation with equally extensive interviews and conversations with people. Regardless of the specific topical focus, fieldworkers usually involve themselves in the daily lives of the people they study, even if only for short periods of time.

Often the researcher focuses on a single, well-chosen community of intermediate size. Where local villages and hamlets contain small numbers of households, it becomes necessary to include several such communities. Research that is concerned with a particular illness of specific population segment (e.g., asthma among small children) requires that the population be large enough to contain an adequate sampling of households with small children experiencing the illness.

Sizes of study samples vary greatly, depending on overall community size, prevalence of specific illnesses studied, the types of data gathered and the resources (including time) available. Nichter and Nichter described a survey carried out in South India, in which a small number of questions concerning food intake during pregnancy, preferred size of baby and relations of food intake to baby size constituted the very simple interview protocol. The simplicity of the interview schedule made it feasible to manage a sample size of 282 participants. Approximately 100 households appears to be a common ballpark figure in much medical anthropology research.

In many cases medical anthropologists find it necessary to collect data from. multiple samples of informants or respondents, in part because of the difficulties in getting adequate representation of the various subgroups in a population. Also, different samples are sometimes selected in order to focus on particular topical areas. In a study of childbirth and obstetrics among the Bariba ethnic group in Benin, Sargent gathered data from several samples:

- 26 postmenopausal women in a rural Bariba village;
- 123 pregnant urban women of the same ethnic group (a clinic-based sample);
- 35 pregnant Bariba women contacted in their homes;
- 50 urban Bariba women currently employed in a cashew factory;
- 77 Bariba women who delivered at the Parakou hospital, whom she interviewed "concerning pregnancy and delivery expenses and hospital experiences".

Thus, Sargent found it important to combine community-based and clinic-based samples in order to get a holistic perspective on childbirth and maternity in the population. When researchers have sufficient co-investigators and research assistants, samples can be larger. Browner and associates carried out research on reproduction and health in a township of 1,800 inhabitants in highland Oaxaca, Mexico.

"In addition to participant observation and intensive interviewing of selected key informants, single interviews were conducted with a 54 percent sample of the municipio's adult women and their husbands. One-hundred eighty women and 126 men were interviewed, with the sample constructed to represent the age, residential and linguistic backgrounds of adult population". Many anthropological studies utilize multiple community samples. In some cases several communities or hamlets must be included for representativeness in a complex population. Gittlesohn, in a study of intrahousehold food distribution patterns, defined his research population in rural Nepal in terms of a network of six villages, in order to include all relevant caste groups.

Quite often the selection of multiple communities is used to operationalize a significant, usually independent, variable. Bentley chose three villages in north India as the population for study of household management of childhood diarrhea, in order to have an experimental and control group. One village was the site of an oral rehydration therapy (ORT) intervention programme and two villages nearby were selected as controls in order to test the efficacy of the ORT programme. In a study of "ethnicity, ecology and mortality in Northwestern Thailand," Kunstadter gathered data from a number of different communities, both highland and lowland. "Community type is the basic unit of comparison. Disaggregation of the population according to type of community shows that fertility and mortality patterns are systematically associated with ethnicity and ecology (location and basic economy).

Populations in the study area allow control of ecological and ethnic variablility by comparing, for example, the same ethnic group in different ecological settings (Northern Thai in Town, Suburb and Lowland Rural communities) and different ethnic groups in the same ecological setting (e.g., Highland Skaw Karen, Po Karen and Lua' with similar swidden economies)". In a similar vein, Hackenberg and associates selected four communities in the Philippines—two sedentary and two migrant groups—for testing specific hypotheses about the effects of migration and modernization on hypertension levels. Gebrian used a multicommunity design in a study of nutrition, rates of immunization and other characteristics in a thirty four-community region in southwestern Haiti. She compared communities in terms of "distance from health centre," "level of community participation," size of population," and other independent variables in testing hypotheses about the effectiveness of primary health care operations.

SAMPLING IN URBAN COMMUNITIES

Medical anthropology in urban sites, particularly in North American settings, very often focuses on one or more ethnic groups within the general population. The selection of such ethnic (or other) subcommunities poses special problems, particularly in identifying the total population of ethnic households from which sampling will occur.

A typical strategy is to identify one or more urban "neighborhoods" thought to be concentrations of the particular ethnic group. In Hartford, Connecticut, Schensul and associates selected two neighborhoods; a public housing project and an area of privately owned houses. After arbitrarily delimiting the two neighborhoods, a system of random sampling was adopted by which 143 households were selected for interviews.

Janes described the difficulties in selecting a sample of Samoan migrants in northern California for his study of hypertension. "The sample selection process involved the following steps: a list of the church membership of two large church congregations was obtained, numbering about 130 households. From this pool, 60 households were chosen at random. This resulted in 89 interviews with men and women in these households. In addition, with the aid of a Samoan research assistant, who was also a well-known and respected member of the community, I selected a sample of 25 individuals from other religious denominations".

The contrast in style between intensive, small-sample research and the collection of survey data in an urban setting is particularly striking in the work of Scrimshaw in the Ecuadorean city of Guayaquil. In the ethnographic phase of research, "sixty-five families in one small area of the squatter settlement were studied for six months using participant observation, conversation, informal interviewing and observation".

An interview schedule was then designed for gathering quantitative data on migration, fertility and induced abortions and was administered to approximately 2,000 households in a squatter settlement and the "central city slum" area, using probability cluster sampling. Scrimshaw demonstrated that her ethnographic sample of fewer than 100 households was in many ways quite similar to the large-scale sample in terms of frequencies of fertility attitudes and behaviors.

BASIC DATA-GATHERING TOOLS

Key Informant Interviewing

Open-ended qualitative interviewing of key informants has long been the foundation stone of anthropological data gathering. Even in projects with a major focus on quantitative survey methods, there is almost always an initial phase of ethnographic exploration—in the form of informal interviewing.Some unstructured conversations and interviews should always be carried out with a variety of informants before more structured quantitative data gathering is initiated.

One major aim of the qualitative research is to develop a good sense of local vocabulary in relation to health problems. The forms of even the most routine questions and observations should be shaped by knowledge of the local language and ecological conditions.Traditionally anthropologists have relied heavily on serendipity in locating key informants.

Loitering about in public places usually leads to contacts with some local persons who happen to feel like talking with the outsider. A somewhat more structured approach is to go from household to household, to visit and to explain the intended research project to individual families.In health care research, whether clinic based or community based, medical anthropologists often follow a sequence of contacts, beginning with local administrators, health authorities and other leaders in the political and administrative hierarchy. Following that sequence of contacts with informants, it has become useful to distinguish three main types of key informants:

- Type 1 key informants: Administrators and officials. The first key informants one contacts are likely to be personnel in government services, police and administratorsin nongovernmental organizations, who am broadly familiar with local situations and activities because of their official dudes. These first-line key informants can help data gatherers to get started in the target community, as they am often gatekeepers who can grant (or refuse to grant) access to clinics, hospitals and community health workers.

- Type 2 key informants: Health workers and community outreach workers. In most areas, both urban and rural, them am outreach workers of health services and other social programs, in government services as well as nongovernment organizations. These informants are particularly important because they can provide direct contacts with people in the target populations. Also, the community outreach workers are often in situations of mediating between service programs and the people's household-based health beliefs and practices.
- Type 3 key informants: Members of the target population. Ethnographic research is rather incomplete until extensive informant interviews have been carried out with members of the target population. In the case of women's and children's illnesses it is commonly expected that the anthropologist visits a number of households, seeking out women who have had a lot of experience with illness management and who am willing to spend considerable time in discussing "cases" and "episodes" of illness. Cam should be taken that these key informants are drawn from several different parts of the community in order to ensure representativeness of the sample, even though key informants are never selected on a random sampling basis.

The numbers of conversations and other contacts with key informants will vary greatly, depending on their availability, their breadth of information and willingness to spend many hours with the interviewer. Typically medical anthropologists, like other ethnographic fieldworkers, rely on a small number of key informants (usually fewer than ten) for a large part of their detailed information, with smaller amounts of contact with their secondary key informants, who are nonetheless very important for cross-checking and triangulation of data.

In some cases anthropologists select samples of informants from different sectors of the local population after carrying out a household census to identify types of households and other variations. Small numbers of representative families (e.g., families with children under five, households with pregnant women, nuclear versus extended households) can then be visited for in-depth ethnographic interviewing.

Ethnographic interviewing and unstructured observations in homes is the crucial stage in many projects, during which the complexities of health care decision making, types of home treatments, attitudes and relationships to health facilities, economic and political issues and many other details of local life are studied extensively, in preparation for carefully designed, structured observations and interviews. The informal contacts with selected households can also include some pretesting of portions of data-gathering formats intended for the more structured portions of the study.

Structured Data Gathering in Samples of Households

Many health-related research projects include several different structured datagathering operations, sometimes using somewhat different samples for the various observations. For example, Bentley, in her research on diarrhea management in north India, interviewed in a random sample of 199 households to collect data on beliefs and knowledge about the causes and prevention of diarrhea and other aspects of explanatory models.

In a later phase of research, mothers of children with diarrhea episodes were interviewed as well as observed in 50 households, to get actual behavioral data. In his research on intrahousehold food distribution in Nepal, Gittelsohn selected six villages within a panchayat, from which he identified a random sample of 115 households.

A number of different interviews and observations were carried out, including a socioeconomic interview, several twenty-four-hour dietary recalls, anthropometric measurements of both children and adults, repeated direct observations of meals (total of 354 meals recorded), collection of morbidity data, in addition to key informant interviews and informal chats with many of the people in the sample. The resulting data set contained twenty-one separate subfiles, totaling 70,000 lines.

COMBINING QUALITATIVE AND QUANTITATIVE DATA

Case 1: Explanatory Models of Illness and Decision Making

One of the more thorough and impressive studies of people's modes of choosing among health care alternatives is that of James Young, in the town of Pichataro in western Mexico. The study is important because it illustrates structured interviewing with small numbers of key informants, after which the methodology shifted to collection of actual illness episodes.

In the first phase of research Young and his wife, Linda Garro, carried out a census of the 509 households in the community and began key informant interviews concerning common illnesses and their characteristics. General ethnographic interviewing was necessary to identify specific types of health care resources utilized by the people, as well as to develop the basic list of locally recognized illnesses concerning which cultural models of explanation and choice making could be derived.

With a provisional set of forty-two physical and behavioral symptoms of illness (each written on a separate card), the researchers asked five literate informants to sort these into piles in terms of severity. The same informants were also asked to sort the illnesses (diagnostic labels) into piles in terms of severity.

In order to get in-depth information concerning the culturally defined characteristics of each of the illnesses, a set of forty-three different illness attributes (questions) were asked of each of the thirty-four illness terms identified in early stages of the research. Ibis time-consuming interview required a number of sessions with each of the ten informants—six women and four men.

The most time-consuming process was the collection of a corpus of actual illnesses, for the testing of the decision model derived from analysis of the data on illness definitions and characteristics. From the original census data, a representative sample was drawn. "Over a six-month period, sixty-two households were visited on an approximate biweekly basis and records were made of each illness occurring among its members". A total of 323 cases were collected.

This research project is very revealing, as it includes a thorough mixing of qualitative and quantified procedures, progressing from pattern identification of illnesses, to the testing of models concerning the ways in the patterns influenced actual behaviors (in the 323 episodes).

Case 2: Modernization and Arterial Blood Pressure

"Hypertension," or more accurately, differential blood pressures, is particularly interesting as a focus of research because the disease is relatively symptom free. As a result, in most indigenous communities there has been no traditional concept. Nonetheless, a great many cultural groups, including many Native American communities, now have cultural models of "hypertension," learned (and modified) from the doctors, nurses and others in cosmopolitan medical services.

Studies of factors affecting differences in blood pressure levels (and outright hypertension) involve use of an overt biomedical measurement (using a sphygmomanometer), plus the observation of cultural, psychological, social and other factors thought to affect arterial blood pressure levels. The following case is important because it exemplifies an increasingly frequent type of study, in which medical anthropologists collaborate with epidemiologists and other biomedical researchers in projects sponsored by international agencies.

Dressler and associates chose an urban area in of Ribeirão

Prêto, a city of 400,000 in Southeast Brazil, for the research. "A variant of cluster sampling was used to draw a sample of 139 individuals. Four broad clusters based on residential and economic sector were chosen and then random samples of 20 households were chosen from each cluster". The clusters that the researchers selected represented agricultural day laborers, continuously employed plantation workers, factory workers and a fourth cluster of bank employees.

For the dependent variable, the research group used the mean values derived from five separate measurements of blood pressure using a DINAMAP Vital Signs Monitor Model 845XT. The automated equipment for measuring blood pressure is much more reliable than ordinary sphymomanometers, as it "virtually eliminates inter-observer variability". Ibis is an example of use of state-of-the-art equipment and methods to manage the biomedical variables in an interdisciplinary project.

Other variables were "index of style of life," composed of ownership of items of material culture and "economic resources," representing occupations of all adult members of the households. From these two measures, an index of "life-style stress" was computed, based on the discrepancies between the two indexes. Dietary data were collected using a series of four twenty-four-hour recalls. Other variables included individual perception of "relative deprivation," "life changes," and the usual age, sex, education and race, as well as height and weight.

The lifestyle index in this study is particularly interesting because it includes ownership of consumer goods (colour television, vehicle, food blender, camera, telephone, etc.) and also items such as yearly trips to Sio Paulo, vacation trips, magazines read per month, newspapers read per week and number of books read per year.

The statistical analysis (multiple regressions) replicated Dressler's previous study of hypertension in St. Lucia, again demonstrating the importance of lifestyle stresses as predictors of differences in blood pressures. In his earlier research, Dressler had carried out extensive qualitative ethnographic research in addition to the quantified data gathering. The Brazilian study, on the other hand, included very little qualitative data gathering.

Case 3: Diarrhea Research of the 1980s

This case departs from the focus on individual projects in order to examine some features of a concerted programme of research on diarrhea that has played a central role in the child survival programs of the 1980s and 1990s, particularly involving Oral rehydration therapy (ORT). Infant and childhood diarrheas have been a leading cause of mortality in most parts of the developing world. Beginning in the 1980s, it became apparent that a major reduction in infant and small child mortality could take place if ORT were regularly used to offset the dehydrating effects of diarrheas, even though ORT is not itself a cure for the illness.

Practically every developing country now has a programme for promotion and dissemination of ORT. In most cases the emphasis is on teaching people to use ORT packets disseminated throughout the primary health care systems; however, some programs have sought to train people to mix home-made oral rehydration solutions, following simple recipes. There has also been promotion of rice-based and other cereal-based ORT. Despite the apparent success in some of these campaigns, most national programs have encountered difficulties in convincing the majority of people to use ORT. And in many instances even the acceptors of the therapeutic regimen do not use ORT in an appropriate manner.

Problems with people's acceptance and proper use of ORT in most developing countries have led to widespread realization of the need for research, including anthropological data gathering, to identify the points of difficulty and to develop ways for improving programs. The study by Bentley in north India is one of a large number of studies by anthropologists during the 1980s. Several of these studies were published in a special edition of Social Science and Medicine.

Biomedical, thinking and directions of research with regard to childhood diarrhea have progressed from a simple focus on ORT, to the examination of complex issues around breast feeding and other dietary behaviors, patterns of use (and misuse) of pharmaceutical remedies and many other culturally mediated beliefs and behaviors. With the realization of complexity has come a greatly increased interest in the possibility that specifically anthropological methods may hold the keys to better management of the diarrhea issue.

A major first step to disentangling the "diarrhea problem" was the shift of focus from the biomedical construct, "diarrhea disease," to the variety of cultural constructions—the emically defined patterns related to diarrhea as illness. One of the more influential studies earlier in the decade was Nations's study of an economically impoverished area of northeast Brazil. She noted the points of incongruity between the prevailing allopathic approaches to diarrhea and the perspectives of the mothers and local healers among the people. She called for new approaches to diarrhea control:

In short, the foremost concern of this alternative approach to diarrheal disease control is to support rather than suppress popular village healing. Doctors must adapt medical terminology to popular usage. They must learn the popular folk explanations for childhood illnesses and explore their relation to biomedical etiologies. Health professionals must also give villagers dietary advice in a way that does not violate harmless food beliefs and that assures the traditional healers power in village medicine. Still, peasant families must also have easy access to effective modern means to save children dying from severe dehydration.

Key informant interviewing to elicit varied terminology for diarrhea, emic types of diarrhea and the exploration of the explanatory models (EM) for diarrhea in given cultures have become core elements of most community-based diarrhea research, even by nonanthropologists.

In north India Bentley found that mothers identified five types of diarrhea: "bloody,": "watery," "bits-and-pieces," "green," and "yellow". Scrimshaw and Hurtado found that explanatory models of diarrhea in Guatenuda included those caused by the mother (due to emotion, physiological condition, etc.). foodrelated diarrhea (hot, cold, "bad," excess), as well as those due to tooth eruption, "fallen fontanel," evil eye, worms and "cold enters the stomach." This complex array of different causes is associated with different choices of therapies.

In a rural area of central Mexico, Martinez asked mothers to sort into groups the various foods that had previously been identified as "foods given to children during diarrhea." The pile-sort task results were then submitted to multidimensional scaling. Martinez asked his community health workers to examine the scaling results and to interpret the dimensions. The general result included the finding that women were differentiating between foods that were appropriate during acute phases of diarrhea and those that are usually fed during the recovery phase.

Most of these ethnographic researchers have used some sort of sampling procedures, but the exploration of folk taxonomies, explanatory models and other aspects of emic views of diarrhea do not generally depend on statistical analysis, other than, at most, frequencies of recognition of taxonomic categories. While some variations are always found in the numbers and types of categories identified by different informants, the range of variation is not usually extensive.

Quantitative survey techniques, on the other hand, are generally used to assess the strength and prevalence of beliefs about causes of diarrhea and especially for assessing frequencies of crucial elements such as cessation of breast feeding, curtailment of solid foods and other potentially harmful behaviors arising from cultural belief systems. Survey techniques have also been used to find out the percentage of people who have heard of ORT, as well as the rates of recent use of this therapy. Coreil and Genece report a survey concerning adoption of ORT among Haitian mothers, carried out in the coastal town of Montrouis and the surrounding villages. As is usual in this kind of study, several weeks of ethnographic investigation were carried out before the survey was initiated. "A random sample of 300 mothers or caretakers of children 0-5 years were interviewed.

Census records allowed us to identify all the 1714 families in the health programme with preschool children"). The researchers made use of two survey teams, "each consisting of 3 interviewers and a supervisor. The project director (author) trained and closely monitored the teams. The 65-item questionnaire was pretested on 15 mothers from an adjacent community". The statistical analysis consisted of a multiple regression to examine the relative strengths of several hypothesized predictors of ORT knowledge and use.

The third major sector of anthropological research on diarrhea has been directed to prospective study of behaviors in the case of actual diarrheal episodes. This methodology requires three main ingredients:

- A representative sample of households containing small children of the requisite age;
- A method for monitoring households so that episodes of diarrhea are quickly identified as they occur;
- A well-designed protocol for direct observations and interviewing concerning the identified cases.

Monitoring and follow-up of diarrheal episodes is time-consuming, as each visit should include direct observation if possible, as well as interviews of caretakers concerning modes of treatment, visits to health care facilities, feeding behaviors, condition of the child and many other details.

As studies of diarrhea have shifted toward possible prevention strategies, interest is developing concerning direct observation of hygiene and sanitation in households, including hand washing, modes of cleaning up after children's diarrhea, maintenance of drinking water and other behaviors. All of the recent studies of actual behaviors have demonstrated that them are wide discrepancies between people's answers to surveys as opposed to actual behaviour as observed directly by the researchers.

At the same time, direct observation of complex health-related behaviors is a relatively underdeveloped aspect of anthropological research and more experimentation is needed to refine the methodology.Collaboration of medical anthropologists with epidemiologists, biomedical researchers and others has been particularly fruitful in the sector of diarrhea control. This is partly because of the widespread realization that qualitative ethnographic work and other anthropological research tools play a vital role in furthering practical understanding of key issues, particularly about the ways in which complex health beliefs, or explanatory models, affect health care decision making.

Case 4: Focused Ethnographic Study of Acute Respiratory Infections in the 1990s

G. Pelto and associates at the World Health Organization (WHO) have recently developed a systematic ethnographic approach to the examination of people's cultural explanatory models of acute respiratory infections (ARI) in very young children. Serious respiratory infection (usually pneumonia) is a major killer of children under five years of age, currently accounting for more than 4 million deaths annually.

In order to promote effective responses to serious ARI among mothers and other caretakers, the WHO planning team developed a set of interrelated data-gathering techniques that produce a relatively clear picture of the explanatory models, as well as behaviors and evaluations associated with the culturally defined ARI illness domain. Here are the main elements in their focused ethnographic study (FES) strategy:

- Data gathering begins with key informant interviewing about respiratory illnesses and their treatments in the local area. The interviews also include a "free-listing exercise" to obtain the vocabulary of signs and symptoms, as well as local, culturally specific names for children's respiratory illnesses.
- Illness episodes of ARI are collected from small samples of mothers in order to assemble more information about behaviors connected with the different categories of signs, symptoms and illnesses.
- Mothers are shown short video clips of sick children (including some with pneumonia) and are asked to identify symptoms and explain "what is wrong" with

the child (what illness does the child have?). The mothers are also asked whether they have perceive the child to exhibit "rapid breathing" (an important symptom of pneumonia).

- The sample of mothers is asked to respond to hypothetical vignettes of children with ARI symptoms. The descriptions are varied systematically to highlight differences in response (recommended actions) for more severe and less severe symptoms.
- The respondents are asked to do simple sorting tasks with sets of index cards, to indicate which specific symptoms are associated with local illness terms. The card sorting is also used to get systematic ratings of severity of the various signs, symptoms and illnesses. Weller (1980) and others have used similar sorting tasks for cognitive mapping of various cultural domains.
- Structured interviewing in the form of paired comparisons is used to elicit the mothers' choices of health care providers. The list of alternatives (doctors, healers, clinics, etc.) is obtained from the key informant interviews.
- Interviews with mothers (and other caretakers) bringing small children to clinics provide more direct information on actual health-seeking behaviors in relation to particular symptoms. A key question in that context is, "Which of the symptoms [that you saw in your child) were the ones that led to your decision to come for health care?"
- Interviews with pharmacists and vendors concerning their prescribing patterns for ARI symptoms and interviews with various indigenous and cosmopolitan practitioners are included for rounding out the picture of the local people's management of ARI episodes. Also, home inventories of herbal remedies, pharmaceutical products and other medicines are obtained from the sample households.

The FES strategy adopted in the WHO ARI programme is a mixture of qualitative and small sample quantitative techniques, many of which are clearly described in Systematic Data Collection, by Weller and Romney. The tools and techniques for cognitive mapping of cultural domains, as well as other basic ethnographic methods for this type of study, are also described in Research Methods in Anthropology.

This example of the use of explicitly ethnographic research methodology at the WHO is part of a rapidly developing trend in international health organizations. There is now widespread interest in adoption of the research approaches of medical anthropology, partly connected with a growing recognition among biomedical personnel and health care planners, of the importance of cultural and social factors in affecting the success of health care programs.

The FES battery of data-gathering techniques is a useful example of the strategy of triangulation in ethnographic research. Data concerning people's recognition and definitions of symptoms and illnesses are collected in illness episodes, free-listing exercises, viewing of the videos and structured sorting tasks. Thus, the data provide systematic cross-checking of both illness definitions and the behaviors associated with them. Also, the authors noted that in their methodology, "Data on antibiotic use patterns are obtained from several FES procedures, including the narratives of past ARI episodes, scenarios, home inventories [of medicines] and interviews in clinics, with pharmacists and with practitioners". Based on those multiple data sources across a number of different research sites, they found that there are widespread problems in misuse of antibiotics: "

- Antibiotics used for mild upper respiratory infections;
- Failure to continue with a full course of antibiotic treatment;
- Saving medication from a partially completed course to use for another episode".

DISCUSSION OF THE CASES

The four cases illustrate a few of the many trends in

contemporary medical anthropology, particularly in fieldwork in primary health care in developing countries. To an increasing extent, medical anthropological research is directed to intensive study of specific sicknesses—either in emic, culturally delimited terms of illness or through study of a biomedically derived disease, for which the relevant cultural explanations, behaviors and other features are explored. Research directed to specific pathological states has the large advantage of delimiting and controlling the range of relevant health behaviors to be studied.

Also it permits the anthropologist to become at least moderately knowledgeable about the relevant biomedical aspects without having to spend months in studying medical textbooks.

Most field research in medical anthropology—in these cases and other studies like them—includes varying mixtures of the following ingredients:

- Initial selection of field site where the sickness condition(s) of interest are prevalent enough to be studied. Quite often the field site is selected because of an ongoing health care programme.
- General, descriptive field research, particularly key informant interviewing, in the area, much like other anthropologists in the first phases of getting acquainted with the local environment and its people.
- Census or enumeration of the local population in order to gather general descriptive information and to acquire the framework for later representative sampling.
- Key informant interviewing and participant observation focused on the particular pathology, to explore "explanatory models," taxonomies, as well as a variety of other areas of cultural knowledge concerning the topical focus.
- Use of pile sorts, triad sorts, sentence frames, or other structured methods with small numbers of informants, in order to refine various aspects of the explanatory models and other aspects of the cultural belief systems.
- Structured direct observations of management of

illness, hygienic-sanitation behaviour and conditions, provider-patient interactions and other behaviors central to the research.

- Structured interviews for eliciting data on main independent and dependent variables from representative samples, in order to test specific hypotheses and to verify major patterns and processes tentatively identified in earlier steps of research.
- Extraction of data from patient records in hospital, clinic, or other health care facility where available. In the usual case, extraction of such data requires permission not only from the administrative personnel who control the records but also from the individual patients or their families. Such records may also include data from special bioclinical observations, including blood pressures, blood samples, urinalysis, x-rays, clinical assessments by doctors and other procedures.
- Analysis of both qualitative and quantitative data, using either micro- or mainframe computers and often both.
- Presentation of results of research and policy recomniendations to persons in the research community and other groups involved in health programme operations.

In addition, medical anthropologists generally collect large amounts of descriptive, contextual information about the community, environmental features, political and economic structures and other relevant material. Most research projects do not, of course, include all of the qualitative and quantified procedures just mentioned. Depending on the time frames of research, the specific questions studied and personnel and funding available, individual projects can range from small-scale studies using one or two of these basic tools, all the way to comprehensive, multiyear programs of data gathering that expand beyond this core list.

ETHNOGRAPHIC ASSESSMENT PROCEDURES

Applied ethnographic research is frequently seen as a desirable first step before health care programs are put into operation. Also, increasing numbers of epidemiologists and other quantitative researchers are realizing the importance of ethnographic research for fine-tuning their approaches to planning of structured interviewing and other research operations. Ilree main reasons for initial ethnographic research are:

- To provide locally relevant cultural information to be used to improve health care programs.
- To provide a baseline of data from which to measure change and effectiveness in such programs.
- To identify locally relevant cultural taxonomies and explanatory models, in order to frame meaningful questions in structured interviewing and observations.

Very often such initial ethnographic research must be completed in a few weeks, so as not to delay the introduction of the health care programme itself However, anthropologists have traditionally resisted what some people have referred to as "quick-and-dirty" applied research.

Based on the general holistic principle common to most sociocultural anthropology, some ethnographers have argued that many months are required just to become familiar with all the relevant cultural features and to become known in a given community setting. Also, learning the local language(s), often thought essential to good ethnographic work, requires a great deal of time. Despite some trepidations concerning rapid ethnographic research, there was a substantial increase in sophistication in systematizing this type of data gathering. In several instances the guidelines for specific data gathering have been set forth in field manuals, particularly when similar data were to be gathered in several sites by different research groups. One of the early examples of such a field manual was prepared by Marchione for the Infant Feeding Practices Study, undertaken by a consortium of researchers from the Population Council, Cornell University and Columbia University School of Public Health.

'Me plan of research called for approximately ten weeks of ethnographic reconnaissance, with a suggested sample of thirty to fifty informants. The informants were to be selected from the same communities in which a later, structured interview survey was to be carried out. A more comprehensive manual for rapid anthropological assessment was designed by Scrimshaw and Hurtado for use in an ambitious programme of research sponsored by the United Nations University, initiated in 1983-1984. Social science researchers in fifteen countries carried out projects that ranged from two or three months to six or eight months, using the draft set of guidelines.

The researchers and their methodological consultants convened in Bellagio, Italy, in 1985 to review the resulting data and to modify aspects of the methodology. The revised set of research guidelines, the RAP Manual was then published for general use in health and nutrition programs. The manual includes an appendix with data collection guides including morbidity history of adult household members, inventory of household remedies, use of health resources, interview with health staff, provider-patient interaction and others.

Scrimshaw and Hurtado commented, "A great deal of practical, diagnostic and applied work can be accomplished in a shorter time and by using a simpler approach". The authors of the manual point out that a great many health care programs have been initiated without any sort of culture-specific map to guide health personnel in adjusting to local belief structures, ecological conditions, economic restraints and other factors affecting peoples' health-seeking behaviors.

Among these several different examples of rapid research techniques, there are several common methodological themes:

- It is commonly assumed that some descriptive materials on the local cultural system(s) are available, so that the researcher does not need to spend time finding out about the economic system, kinship and social organization and other general features.
- Familiarity with the local language on the part of the researchers, or else use of local research workers as interviewers, is generally assumed.

- The extensive version of the holistic assumption is rejected in favour of a more limited style of multifactor research. For example, in research on diarrhea or acute respiratory infection, it is assumed that data gathering can focus very specifically on the illness itself, plus a clearly specified list of contextual factors. Very little general ethnography is needed.
- The specific ethnographic data to be collected are thought to require small numbers of informants. Samples of thirty to forty respondents are usually sufficient to establish the local vocabulary, explanatory models and other patterns.
- When research is limited to small numbers of informants, contacted during a fairly short time period, considerable care is usually exerted to ensure representativeness of the sample, in terms of local subgroups, age and sex distribution, socioeconomic status and other dimensions of variation.
- The use of focus group interviews or group discussion sessions is commonly used in the exploratory phases of the research.
- Inferential statistical analysis is seldom appropriate in the rapid ethnography methodology, but descriptions can include presentation of frequencies of responses in various categories.
- The rapid methodologies sometimes include limited use of survey methods near the end of the data-gathering process.

The rapid methodologies have developed primarily in response to the requirements of applied primary health care programs. In many cases the rapid ethnographic assessment is needed to produce basic data for designing health care intervention programs. However, the Rapid Assessment Procedures manual by Scrimshaw and Hurtado was developed in relation to evaluation of ongoing nutrition and health care programs.

Chapter 8

Biocatalysts

In 1812 Kirchhof discovered that if starch were heated with sulphuric or other strong acid solution it was converted into sugar without any change or loss in the acid itself. Thénard noticed that hydrogen peroxide solution decomposed, giving off oxygen in the presence of undissolved bodies, such as finely divided platinum, silver, manganese dioxide or blood fibrin, not entering into the reaction, whose mere presence was sufficient to produce this effect. Sir Humphry Davy showed that platinum in the powdered or spongy state, and therefore of greatly increased surface area, can oxidize alcohol vapour exposed to the air without the application of heat.

Berzelius in 1837 called general attention to the meaning of these facts: "In reactions of this nature there is present a chemical force quite distinct from any hitherto known," and he proposed to call such operations catalytic in view of their being absolutely different from analytic actions resulting from the use of those reagents normally used in chemical research. In ordinary analytical processes the reagents come in according to the laws of affinity of normal chemical relationships, whereby they are transformed whenever they are present. On the other hand, in catalysis, the factors giving rise to the action have their effect merely by being present and in themselves suffer no intrinsic change whatever.

Prior to this basic statement by Berzelius, which is one of the cornerstones in the history of chemical science, Dubrunfaut in 1830 had already made one important discovery, namely that fermented barley extract or malt converts starch into glucose even in the cold state, similarly to the way an acid acts in the presence of heat as above mentioned. Shortly thereafter, Payen and Persoz in 1832 succeeded in separating the amylolytic principle from malt by extraction in water and precipitation with alcohol, giving it the name diastase.

They considered this agent identical with the natural ferments, which cause souring in wine or milk. Robiquet consequently showed that bitter almonds contain a protein substance which breaks amygdalin up into prussic acid and glucose; this active protein was christened emulsin by Liebig and Whöler.

The fermentative properties of the gastric and pancreatic juices came under consideration practically at the same time. Pepsin was discovered by Eberlé and Schwann, and trypsin by Corvisart. These newfound substances were to be of the greatest value in bringing the study of digestive chemistry nearer to that of the catalysts. Thus it is that the digestive diastases at that time came to be the most typical examples of physiological catalysis, and that is how diastasis and catalysis acquired their greater renown as factors in the vital processes. The idea got abroad that certain animal organisms contained or produced diastases in just the same manner as did certain vegetable tisses.

FERMENTATIONS

Pasteur's researches on the processes of fermentation extended from 1857 until 1870. Some score years previously, in 1838, Cagniard de La Tour asserted that fermentation was due to microscopic beings. Berzelius opposed this idea that fermentation should be one of the vital processes. "The case of fermentation," he wrote, "is similar to that of the decomposition of hydrogen peroxide in the presence of platinum, silver or fibrin. A catalytic action like any other."

Pasteur's great work definitely proved the effect of various yeasts and bacteria in different fermentation processes and determined the conditions in which such processes could develop. "Fermentation is the work of living organisms," he wrote, and very soon proved the fact that not only are single-celled micro-organisms possible factors of fermentative change, but also that cells from metazoic or metaphytic tissues existing communally in polyplastid organisms might also play their part in the process. Fermentation might therefore be shown to be a vital property inseparable from the structure and functions of the cell.

Lebig, from 1839 onwards, maintained that ferment is a soluble matter of protein nature working through catalytic action, and that structural or functional life was not essential to demonstrate its effects and properties. Ferment, he said, was neither an organized nor an organic product with all the complexities which organized life would entail. Liebig's "soluble ferment" was the same as Payen and Persoz's "diastase" — later on Duclaux was to propose for general acceptance the generic use of the word diastase to distinguish soluble ferments — it is the same thing as Kühne's "enzyme" or "zymase," etc.

DIASTASES AND PHYSIOLOGY

The labours of Traube (1838), Berthelot (1860), and Hoppe-Seyler (1876) finally facilitated the correlation of these various theories, whether life or mere catalysts, were responsible for fermentation. Numerous diastases were discovered formed from the most diverse types of cell. That is how cells make diastases and how the fermentative effects of such cells become evident through the diastases. Life is, therefore, catalytic through the action of its diastases originating from vital actions and comes about at the same time through catalysis; which suggestion does not, however, prevent the parallel existence of catalysts not endowed with life and even of an inorganic nature.

These concepts found definite confirmation as a result of the researches made by Buchner and published in 1897; yeast extracts totally free from cellular matter even in fragmentary form, containing therefore only dissolved matter formed from living cells and separated entirely from cell plasm, were capable of developing alcoholic fermentation if they came into contact with glucose. The diastases of alcoholic fermentation are manufactured by yeast cells, that is living organisms, in the same way as pepsin, trypsinogen, amylase or lipase are manufactured by the gland cells of the stomach or pancreas. Diastases are therefore merely products of metabolic action or cell functioning. These ideas became still more positive when the intervention of the diastases was speedily discovered to be a factor in all nutritional processes and an influence in the development of all functions. Diastases became accordingly something more than mere digestive factors. Wittich in 1873 proved the diastatic nature of the evolution of glycogen in glucose in the liver.

He isolated the diastase by solution in water and glycerine and precipitation with alcohol. Claude Bernard in 1877 after having many doubts on the matter, confirmed Chauveau's assumption of 1856 that glucose was consumed by the tissues. Similar consumption occurs in the blood, which when protected from putrefactio n at room temperature in the laboratory rapidly loses its sugar content. Lepine in 1890 and Arthus five years later attributed the glycolytic diastase or diastases present in the blood to be of leucocytic origin, but immediately pointed out also that the tissues in general and the muscle in particular produced similar diastases in greater quantity. Starting from this point, discoveries of numerous and distinct diastases from all the tissues came thick and fast. Each act of metabolism arises from the action of the corresponding diastase or diastatic group, and as time goes on we are daily increasing our knowledge of fresh details in regard to the intimate relationship between nutritional reconversion and the functioning of the organs.

The study of diastases thus covers a very wide field, particularly during recent times. As Fearon said in 1940: "Since 1900, research on the diastases has reached such a stage of development that each single diastase or a single feature of a particular class of diastase is sufficient to occupy fully an entire group of research workers."

THE NATURE OF CATALYSIS

Ostwald, in 1896, defined catalysis as an acceleration of the corresponding chemical reaction without modification of the state of balance of such reaction and without affecting therefore the amount of energy liberated or absorbed thereby. From that time on, researches have been repeatedly started for the purpose of investigating the nature and course of diastatic actions.

Among the many scientists concerned, we may mention Jacobus Henricus van 't Hoff, Arrhenius, Armstrong, Duclaux, Sabatier, Födor and others who busied themselves with the question of diastatic catalysis and conditions leading thereto. They were able to assert once more that diastasis is one particular case of catalysis.

According to Hughes a diastase is any substance which by its mere presence allows certain molecules to undergo chemical change through the effect of a critical quantity of energy less than what would be necessary were the catalyst not present.

There are many types of catalysts at work in living matter; sometimes they are the hydrogen, hydroxyl or iodine ions and their like, at others the catalyser may be a dissolved organic particle, metabolite, hormone or vitamin, an inorganic colloid like the metallic suspensoids or lyophobic rolloids, distinct organic particles, or diastases of some special type; and perhaps, finally, combinations energetic in action rather than of a specific material effect, for instance, porous walls (Becquerel, Giraud), radiations (Victor Henri, Achalme), or membranes or other cellular structures.

THE BIOLOGICAL SIGNIFICANCE OF CATALYSIS

It is through catalysis that vital phenomena are developed in living organisms, since only by this means can one explain how otherwise impossible chemical reactions take place. There is no single physiological process which is not under the effect of many vital catalysts.

Diastases, hormones, parahormones, harmosones, vitamins and certain ions govern the activity of the organism as a whole as well as each of its parts, and control the appearance and development of the most diverse phenomena of life.

The vital catalysts are countless in number and act in a systematic manner co-ordinated rigorously within the framework of the being, their intrinsic potentiality being most intense. Haldane for instance has estimated that a molecule of catalase can decompose somewhere about 200,000 hydrogen peroxide molecules per second. Conversely, each molecule of the substrate will come into contact with a large number of diastase molecules and thus be converted. Haldane's calculations show that the average atom undergoing oxidation in the course of its metabolic life through the cell encounters no less than 100 catalysing molecules one after the other.

These facts show that the immense metabolic activity present in the living state can be explained by the presence and working of suitable catalysts. It is not possible to guarantee that these figures completely represent the true state of affairs, but, even though they be mere approximations, the extent and intensity of the biocatalytic function can thus be inferred.

Such catalysts must unfailingly be present in all living bodies, whether plant or animal, and it is because of them that life continues to exist. Without catalysts there would be no metabolism, and without metabolism there would be neither function nor morphogenesis. It is impossible to conceive of any vital manifestation of any type without the assistance of the relative catalysts.

CATALYSTS AND LIVING CONDITIONS

Catalysts enter into the appearance and development of the vital manifestations. Conditions most favourable to them occur naturally in circumstances which are likewise best for the maintenance of life: moisture, for instance, since biological catalysts must be dissolved in water — the substance cannot act unless dissolved; warmth, since they act best and with greater intensity at the body temperature of warmblooded creatures; particular osmotic pressure and a predetermined acid-base balance individual to each one of the various groups of diastases; ionic surroundings and the presence of certain substances also help.

In this way complex diastases may be observed, and the

functions of co-diastases, activating or inhibitive agents, and antidiastases must also be borne in mind. Diastases may have their source in the inactive form of prediastases and then be set in motion by various means.

Other vital catalysts, like the hormones or vitamins, demand the help of particular circumstances, or contrariwise find their physiological effects opposed by the presence of particular types of disturbing factors. Various metabolites, of which we could mention many, such as glutathione, diphenyloxylic acids, adenylic acid, flavin, etc., exert powers of fermentation analogous to those of the diastases, probably without any diastase whatever having to intervene. As may be seen, the catalysing agents in organic life are of a varied character, the wealth of their activity belongs to life itself and the need thereof is equally intrinsic to requirements for the continuance of life.

In little more than a single century immense progress has been made in theory, particularly as far as concerns biological catalysis and the real importance of those phenomena, originally thought to be nothing more than individual simple applications of organic chemistry of no greater interest than, say, their value in industrial application; the fermentation processes, for instance, have been effectively developed. It is obvious that a lot more still remains to be done; catalyst mechanisms have been shown to be indivisible and of the same essence as all the processes of life. "Life is ferment," as Paracelsus wrote in the sixteenth century.

A DIFFERENT FORCE

It is indeed surprising that the blood-stream supplying all organs alike should, without the help of any other liquid whatsoever, produce saliva, milk, urine, and so forth. In 1812 Kirchhof made a discovery which gave us a first glimpse at the life processes though remaining still far from a full explanation thereof. Kirchhof observed that starch could be converted into glucose by the action of dilute sulphuric acid, the latter remaining unchanged and unconsumed in the process. Thénard showed subsequently that hydrogen peroxide can be decomposed by finely divided platinum and other substances in suspension. Platinum itself in the spongy form is capable of oxidizing alcohol at room temperature.

It is common knowledge that the conversion, for instance, of sugar into carbon dioxide and alcohol takes place during fermentation under the action of an insoluble body which we call ferment, and that animal fibrin, coagulated albumin obtained from various plants, cheese and similar substances behave like such ferment though with less intensive effect. Such reactions cannot be explained by the laws of chemistry, as a double decomposition taking place between the ferment and the sugar or any similar reversible reaction.

It may perhaps be better to compare such conversions with inorganic processes such as the wellknown decomposition of hydrogen peroxide in the presence of platinum, silver, fibrin, and so on, from which we must assume that the action of the ferment may be of similar character. I do not by any means imagine that we are dealing with chemical forces different from those exhibiting themselves as properties of matter. In some cases it is quite possible to explain reactions of the type under consideration by availing ourselves of the well-known laws of chemistry and physics.

Therefore I propose calling the power of substances causing the conversions above mentioned under the conditions indicated catalytic, contrasting the word catalysis with the term analysis, so much in vogue for signifying the separation of the components of a body by means of chemical affinity in the normal sense.

Catalytic power seems to consist in the faculty certain substances exhibit for setting in action potential affinities and chemical activities which may be lying dormant at a particular temperature and are presumably something quite different from the real affinity between the bodies, since their action occurs through their mere presence.

There are well-justified reasons to suppose that in living animals and plants, thousands of catalytic processes take place between the organic fluids and tissues, whence we get the formation of countless heterogeneous chemical compounds, starting with plant saps or animal blood whose composition does not change through the fact that so many and various products are manufactured. It is likely that some time in the future we shall find out the cause of it all is the catalytic power of the tissues forming the organs of the living body. Every solid living tissue placed in contact with a liquid seems capable of acting chemically upon the substance in solution, whilst at the same time the dissolved matter can react without any interchange of elements. Catalysis is both chemical and physical in its nature.

Attention has recently been called to the fact that a body may exert influence by mere contact upon some other body which is thereby decomposed or combined. Platinum, for instance, does not decompose nitric acid neither at the boiling point of the acid nor in the finely divided non-reflecting state (spongy platinum), but an alloy of silver and platinum is easily dissolved in nitric acid. The oxidation suffered by silver gives rise to a similar change in platinum. In other words, platinum in the presence of silver undergoing oxidation, acquires the property of decomposing nitric acid. Copper does not decompose water even when it is treated with boiling dilute sulphuric acid, but an alloy of copper, zinc and nickel dissolves under such conditions, hydrogen gas being liberated.

These and similar reactions cannot be attributed to "special forces" differing from those governing chemical affinities, forces which would be set in motion by a touch as light as a feather. The unstable substances which are so easily decomposed give the greatest number of instances of this type of phenomenon. The components of these bodies are found in such a state of tension that the slightest modification of their atomic balance overcomes the effects of chemical affinity. They exist only by the force of inertia and the slightest shock or disturbance is sufficient to destroy the mutual attraction of their components.

Hydrogen peroxide belongs to this class of substance; it is decomposed by any product capable of attracting the oxygen which goes into its composition and even by the mere presence of certain substances such as finely divided platinum or silver, which themselves do not enter into combination with any of its constituents.

The cause of the sudden separation of the elements in hydrogen peroxide has been held to be quite distinct from that giving rise to a common chemical decomposition, and has been called the "catalytic force." It should be remembered, however, that the presence of platinum or silver has no other effect than to accelerate the decomposition of the hydrogen peroxide, which would take place even without the intervention of those metals but at a very much slower rate.

Certain organic bodies seem to enter spontaneously into a state of fermentation or putrefaction, especially those containing nitrogen. It is worthy of note that very small quantities of such substances in that state possess the power of bringing about similar transformations in unlimited quantities of substrate. Thus a small sample of grape-must in fermentation added to a large amount of the same liquid in the still state causes the whole lot to ferment. Similarly the smallest amount of putrefying milk, meat extract or paste added to fresh milk or meat will soon cause the latter to putrefy in the mass.

All these changes differ apparently from those ordinary decompositions which are developed through the effect of chemical affinity. These are chemical actions, conversions or decompositions which become stimulated by substances in the same condition. It is demonstrable that inorganic compounds made up of the simplest molecules are the more stable and resist decomposition, whilst others more complex in structure are more liable to decompose.

The cause seems clear: the greater the number of atoms forming a molecule, the greater the number of ways in which the said atoms are mutually drawn together. Such complex balances are practically unstable and persist only through inertia. In this way, complicated chemical combinations, like the sugars, for instance, become the object of such decompositions or fermentation.

We have seen that some metals acquire the power, not inherent in them, of decomposing certain bodies, such as water or nitric acid among others, through mere contact with other metals in the act of chemical reaction. We have also seen that hydrogen peroxide and other peroxides which decompose easily can transmit this quality to other substances present. This is the same thing that happens, as we may also recall, in the case of fermentation or putrefaction brought about by organic matter which itself is fermenting or putrefying.

Different bodies, in the act of combining or decomposing are found to be in a position to induce such processes, that is to cause changes in balance between the component atoms of the molecules, particularly of organic molecules which are highly complex, and causing these atoms to become arranged in other states of balance of greater stability, i.e. in the form of smaller moleculed bodies, in accordance with their natural affinities. Among the substances especially active in this respect must be reckoned in the first place those organic substances containing nitrogen, particularly the albumins.

Chapter 9

Growth Metabolism

Growth in the animal and vegetable kingdoms is a process from which no individual is exempt. Growth, that is increase in size and weight of the living matter forming the individual, results from assimilation. By assimilation, matter similar to that of the organism itself is built up from the foods consumed, which latter are moulded and converted to form the specific substance of which the whole individual is composed. Despite their general similarity of composition, both the living substance and the greater portion of the organic material in various organisms are different and distinctive, certainly for each particular species and even for each individual within any particular species.

The basic property of five matter is its power of building up substances similar to itself. Consequently, synthesis of living substances always presupposes the influence of some other five matter, and synthesis can only take place when such matter is present. Synthesis comes about in such a manner that fresh living matter makes its appearance complete with chemical, histological and anatomical formations identical to those of the organism entering directly and immediately into the assimilative process.

In this identity between the products of assimilatory synthesis and the organized substances which are the instruments of the synthetic process, consideration must be paid to the specific assimilation applying to each organism, to each cell thereof, and even to each differentiable portion of such cells. Thus, the growth of an individual with all its chemical and morphological characteristics comes about through progressive assimilation and the number of individuals forming a species is likewise increased. Assimilation is the process of manufacturing living matter proper, that is to say, the essential cell plasma, residing and living within the cell, whilst at the same time it produces inclusions, paraplastic materials, nutritional reserves or functional products playing a special part in metabolism and in the body functions. In short, through assimilation, nature is able simultaneously to create both the machine and its fuel. These assimilative and anabolic operations are of course endothernmic, whence it happens that assimilation must integrate not only matter by itself but energy as well. Typical of assimilative changes are those brought about by the action of chlorophyll.

It is known that this substance causes water to combine with carbon to form one of those self-same paraplastic substances just referred to, in this case starch, which forms a food reserve capable of subsequent use for countering the many necessities of the individual. To the substance, in this instance inorganic, namely water and carbon, energy in the form of sunlight has to be added so that organic matter may be synthetized. Organic molecules, being much bigger and more complex than inorganic particles, contain more heat than the latter.

The greater the complexity of the number of atoms forming them, the more the energy within the molecules. Assitmilation, the combined whole of the synthetic process, demands fixation of the energy supplied to the material elements making up the large particles which, in the first instance, are organic, and then become alive. It must therefore be borne well in mind that assimilation does not consist merely of the manufacture of complex material edifices or particles of vast structural complexity, but that it must also supply such systems with the amount of energy necessary for their synthesis.

ASSIMILATIVE DISASSIMILATION

The source of this energy is sunlight. Once the glycides

are produced, through union of bodies low in energy such as water and carbon, such glycide reserves will be able to supply the energy necessary for further assimilative changes because of their disassimilative decomposition. Energy will pass from a system in disassimilation to another in synthesis, as shown by Pasteur in 1875, when he wrote: "The energy supplied by oxidations typical of disassimilative operations can be used for completing synthesis within the organism." Therefore assimilation and disassimilation develop in indissoluble conjunction one with the other.

Coupled catabolic and anabolic reactions frequently occur as instanced by Meyerhof in an interesting. example, namely, the resynthesis of glycogen from lactic acid in the muscle, by the use of the energy liberated from a portion of the lactic acid. Such energy of chemical origin being applied to the unconsumed portion of the lactic acid raises it by assimilative operation to a glycide. Such processes are now known as "the Pasteur-Meyerhof reactions," and occur repeatedly in animal and plant life. Plants, besides being able to convert carbohydrates into fats, also manufacture proteins by incorporation of nitrogen with ternary molecular structures.

This stands for synthesis and by the same token a need for energy, which is introduced by decomposition, most certainly oxidation, of a fraction of the glycides resulting from the chlorophyll action. In animals too, the immediate organic principles making up the foodstuffs have to rise to the level of more complex living matter with a greater heat content than the feeding matter itself.

This is effected by consuming a part of the food components in disassimilative operations in such a way that they supply the energy necessary for such assimilation. In fact, the three main objects of dynamogenic feeding within the limits of animal metabolism are: maintenance of kinetic energy, generation of heat, and output of energy necessary for assimilatory synthesis.

FUNCTIONAL DISASSIMILATION

Once living matter takes shape, once the peak which is

the giant living molecule has been reached, that disassimilative exothermic disintegration commences whereby the vital processes are sustained. The functions get their energetic nutriment from these disintegratory processes which always finish up with a credit balance.

The consumption of dynamogenic matter does not take place solely at the expense of living matter properly so-called, as on the whole it uses up reserves stored within the cells, forming the fuel referred to above. Living organisms, from this point of view, may be compared to a heat engine burning a special type of fuel but also partly consuming certain portions of the engine itself, which portions are immediately replaced. The flow of energy in the case of live organisms cannot be compared to the current of a river as seen from the banks unless we take it that a portion of the banks themselves are carried away by and in the stream itself at the same time.

LIVING MOLECULES IN THE STATIONARY STATE

Metabolism, an essential feature of life, is a continual self-destruction and self-recomposition of living matter. Cuvier called it a "living whirlpool." Goethe before him had written "life is a flame," whilst, according to Buffon, it was "a minotaur devouring its own substance." It is in fact a constant flow of matter and energy; living systems are complexes in a stationary condition (Ostwald), seat of continuous changes of matter and energy under the cloak of stability.

Pflcüger in 1898 portrayed the living molecule as a figure in threedimensional space made up of rays darting upwards, downwards, rightwise and leftwise, forward and backwards from a common centre reaching towards the surface of an ideal sphere, such rays being composed of groups of radical atoms corresponding to the various immediate principles, proteins, glycides and lipids forming an extremely complicated particle which has a very high molecular weight despite the low atomic weights of its separate elements.

To this concept of the living molecule Verworn gave the name biogen. From the biogen linear atomic groups of greater or less complexity, proteins, carbohydrates or fats, or portions of such radicals, emerge into the open, these being the "lateral chains" which Ehrlich in 1904 sought to use as an explanation of the phenomena of immunity.

These lines of atoms ceaselessly falling from the biogen into the disassimilative processes would leave atomic remnants in the remainder continuum of the biogen, upon which by assimilation further chains would build up similar to those already broken away. We must imagine the biogen, therefore, as the seat of constant renewal of its constituents, some disappearing whilst others make their appearance, and that, despite all this, the chemical identity of the whole is preserved. There takes place in it — and this is the essence of life — an uninterrupted flow of matter and of energy; assimilation makes up for losses caused by disassimilation. Disassimilation liberates energy whilst assimilation on the contrary demands it. Thus it is that the stream of matter in metabolism entails a corresponding flow of energy.

Bertalanffy names the aforesaid uninterruptedly continuous physico-chemical vital process an "open system." The organism is a system or combination of unified systems in an open steady state of rhythmic constancy, to which the living being tends ever inexorably to return whenever disturbances become apparent. This rhythmic recurrent balance is a basic characteristic of existence, and when it is interrupted, life ceases.

Biogens of various types go into the formation of cell plasma, cells, tissues and organs. Biogens are arranged according to their physicochemical structure, and consequently result in cytological and histological structures which make up the complete anatomy. Nutrition is a property of biogens and therefore of the cells. Living matter attains its animate condition through its nutritive capacity; metabolism must be therefore one of the principal factors in life.

Gautier wrote in 1892.: "In the structure and organization of chemical molecules forming the protoplasm and in their manner of association one with the other lies the fountain-head of the whole range of elementary phenomena of life. Variation of the integrating molecule causes changes also in the style of reaction throughout the whole organism; vital processes are the direct consequence of the chemical functions of the constituent particles."

METABOLIC MECHANISMS

Assimilation and disassimilation are so bound up with one another as to form a single feature. The former was a subject of research long before the latter. Knowledge of the existence of certain processes of assimilation goes right back to ancient times, whilst disassimilation is on the other hand a comparatively recent discovery. Nevertheless greater attention has been paid to the study of disassirailative processes than to acts of assimilation. Both these processes are equally important in forming the metabolic whole. They are of equal effect in the state of nutritional balance; anabolism overrules catabolism during growth, and conversely, disassimilation preponderates in all the aspects of orgainc impoverishment.

Disassimilative changes present nothing which is exclusive and particular to living bodies. As we know, Lavoisier taught that breathing is exactly the same thing as combustion. Hydrogen and carbon are consumed inside living organisms in exactly the same way as in any other combustion process. Living matter disintegrates and decomposes into smaller particles during disassimilation as it does in death. Disassimilation is like bodily decomposition. In both cases the breaking down operations are exothermic and in disassimilation there resides a store of energy available for the functions.

Thus Claude Bernard was able to say: life, that is function, is death. Catabolism, like putrefaction, is an act of decomposition, simplification, a tendency to change back living or organic matter into inorganic matter. Such changes presuppose a positive thermal set-up both inside and outside the organism.

Assimilation, on the contrary, occurs solely in living organisms and is specific to them as we have already seen. Le Dantec in 1904 catalogued the characteristics of the disassimilative processes as compared with the assimilative, putting them into the form of two basic formulae of nutrition. In disassimilation, which in general circumstances hardly differs from any other disintegrative or analytical change, the substances reacting in the first term of the equation are not to be found in the second, a total conversion of substance having taken place both in the nutritive material and in the cell components.

$$N + O = R$$

i.e. Nutriment plus Oxygen equals Residues.

In the assimilation formula, one of the reacting terms, the living matter, is found generally in increased amount in the second term:

$$N + L = nL. + R$$

i.e. Nutriment plus Living matter equals n times the Living matter plus Relsidues.

The object of assimilation is to build up living matter by consumption of foodstuffs. That is why, therefore, foods are divided into respiratory-as they were called in former times-or dynamogenic, and plastic or morphogenic; in other words, foods for heat-generating combustion and foods for the formation of living matter. These two facets of metabolism would not occur independently. Claude Bernard himself pointed out that prior disassimilation is a factor of assimilation.

To repair a breaking off of portions from the biogenic particle, obviously such amputation must first have taken place. As we have shown, it might be thought there was some contradiction between assimilation and function, but elementary observation proves the contrary, namely, that an organ in operation continues to develop, as for instance a muscle. Rodrigo Lavin in 1900 emphasized this when he asserted that assimilatory synthesis is a functional phenomenon. In a function, of which the muscle is the most characteristic example, food reserves are consumed mainly as a supply of fuel for the production of energy.

Living matter as such takes part in the chemical changes demanded by the function and that entails an increased biogenic disintegration. The replacement of molecular complexes destroyed by the functions is more than taken care of by reparatory mechanisms, from which we get an increase in the synthetic operations, and an augmentation of the living matter accompanying and distinguishing the function.

ANOXYBIOSIS AND OXIDATION

Disassimilative disintegration was considered a few years ago to be exclusively of an oxidizing character, according to the ideas expressed by Lavoisier. Our increased knowledge of facts now shows that the chief part in disassimilation is taken by anaerobic operations without the aid of oxygen, these being the chemical reactions labelled by Pasteur as fermentative. Oxygen then acts upon the products of anoxybioses whilst combustion completes the acts of disassimilation. This twofold aspect of the process was brought to light mainly through Gautier's researches in 1898 and has more recently met with repeated confirmation.

Disassimilation takes place through the agency of various diastases leading the living matter or reserves through all stages to excretion. Assimilation is similarly the work of synthetizing diastases, i.e. enzymes, whose exact operation is at the present time even less known than that of the catabolizing diastases. Each of these two types of diastase regulates its activities in such a way as to ensure exact adjustment of the intensity and quality of nutritional changes to the vital needs of each individual, be it plant or animal.

The result of this is that the intensity of nutritional changes comes about neither through chance nor even by reason of special substances at the disposal of the living matter in its uninterrupted course of metabolism. Living matter automatically regulates the intensity of chemical changes taking place in it, by means of properties which we must perforce call vital, intrinsic in itself, from the very fact of its being living matter.

This impelled Pflüger to say in 1905: "The live body cannot be compared to a fire which burns all the more fiercely the more fuel is thrown on to it; quite the contrary: whatever may be the quantity of material, it oxidizes only the exact amount needed for its functional needs.

If the materials of combustion are excessive in amount, they stay stored up in the form of fat. Consequently, what determines the total of work in our organs is not the amount of matter supplied by the food, but the quantity of material actually used." To which Bouchard added in 1906: "A portion of body albumin, eight parts in a thousand in young people and four in a thousand in older folk, is probably destroyed every day and replaced by an equal amount of albumin derived from the food consumed.

The amount of living albumin renewed comes out practically constant for each stage in life and for each set of circumstances. It varies according to age, decreasing as time rolls on; it increases during certain illnesses particularly when fever is present and decreases in others, especially in certain diatheses. It is not affected by muscular effort nor by cold, since the foodstuffs supply — quite apart from the albumin which has to replace that lost in the tissues — sufficient organic matter to liberate the calories required by the effort or by the struggle against cold.

Neither is it affected by variations in the intake of oxygen or protein foods, since such foods are ingested in sufficient quantity."Nutritional changes develop therefore in all forms of animal and plant life as the basis of some form of vital manifestation, animal or vegetable. Disassimilation is particularly manifest in animals, which move and, on getting heated up, require calories which are in fact supplied by the disassimilative, exothermic operations. Apart from such differences in quantity, the metabolic mechanisms in animals and plants — not counting the chlorophyll action—are identical.

CLAUDE BERNARD

If in regard to inorganic matter it be admitted and rightly so that nothing is lost and nothing is created, nevertheless in the case of living beings it is quite another matter. Everything is created morphologically, becomes organized, and then dies or is destroyed. In the developing embryo, muscles, nerves, bones, etc., appear in the spot corresponding to their place in the organism repeating the form of the individual from which the egg was generated.

The surrounding matter is assimilated to the tissues either as a nutritive principle or as an essential component. Thus is created the organ which can be distinguished by its structure, shape and properties.

On the other hand, the organs are destroyed and disorganized continuously and precisely through their own functioning. Such disorganization constitutes the second phase of the great drama of life. The first of these two orders of phenomenon is individual and special to the living body. This evolutionary synthesis is the very essence of life. The second order of phenomenon, the destruction of life, is of a physico-chemical nature, normally the result of combustion, fermentation, putrefaction or some other action, comparable in short to a large number of chemical acts of decomposition or disintegration.

These are true phenomena of death when they take place in an organic being. And it is worthy of mention that we delude ourselves habitually in such form that when we want to point out phenomena of life, what we really talk about are phenomena of death. Vital phenomena properly so-called do not attract our attention. Organizatory synthesis works silently within the body, its actual occurrence well hidden away from sight, gathering up secretly the materials which will afterwards be consumed.

We take no account directly of these phenomena of organization and assimilation. Only the histologist or embryologist studying the development of tissues, cell elements or the living body as a whole becomes aware of the changes and phases which reveal this hidden working; in one case it may be the deposit of material, somewhere else it will be the formation of a skin or a nucleus, further on still a multiplication of cells or a renewal of some organ or other. The phenomena of destruction or death and disassimilation are the most obvious to our senses and seem therefore to us to characterize life.

Their signs are clear; a movement takes place because a

muscle must have contracted as a result of some corresponding wish; when stimulated a gland secretes. It is the substance of the muscle, nerve, brain or gland which is disorganized, destroyed and consumed.

The existence of all animals and plants is maintained by these two types of necessary and inseparable processes: organization and disorganization. Our science has to fix the terms and conditions which guarantee the correspondence of these two types of phenomenon. This division of vital phenomena which we have adopted — assimilative and disassimilative — is an exact expression of the truth and comes about simply from proper observation of life phenomena. But life is the combination of both types of manifestation.

CHARLES BOUCHARD

A common property of all living particles is a particular molecular movement which is not observable in either inorganic or dead matter. The effect of this movement is to introduce into the living particle certain extraneous non-living substances and submit them to chemical metamorphoses which I will call vivifying, since by means of such changes the said substances come to form a constituent part of the live element and participate in its life; there are also other changes which I would call retrograde, whereby living matter ceases to exist and is converted into products of decomposition which have to be expelled from the organism.

The distinctive feature of the activity of any living particle is, therefore, a twofold continuous molecular movement of introduction and expulsion simultaneous with a double operation — also continuous — of chemical transmutation, the one following the introduction of substance and the other preceding its ejection. On the one hand we have translation, and on the other transmutation: a double physical operation simultaneous with a double chemical effect.

If nutritional material conversions are a common characteristic of anything alive, if nutrition is therefore the general characteristic of life, the best definition of life itself will be that given by Aristotle: "Life is the combination of the operations of nutrition, growth and destruction." This idea corresponds with what Blainville set out in other terms: "Life is the double internal motion of composition and decomposition which at the same time is general and continuous."

Nutrition therefore — we may repeat — is life; it is life with its double motion of assimilation and disassimilation, of creation and destruction at the same time. Assimilation and disassimilation are in general parallel and simultaneous but not invariably so, not even in the physiological state. Growth is the predominance of assimilation over disassimilation. In pathological conditions this predominance is the cause of over-development or hypertrophy, whilst atrophy and cachexia come about through excessive disassimilation.

And we may affirm withal, that even when the physical and chemical actions — each of them twofold in the sense of being integrative and disintegrative — are not found to be indissolubly bound together between themselves at every particular moment, they are jointly and severally equally necessary for the maintenance of life. None of them may be hampered or suppressed without the nutritional molecular motion being held up and life being consequently extinguished.

ARMAND GAUTIER

Life makes its appearance in the cell and passes from generation to generation thanks to specific substances transmitted from parent to offspring and determining the organic forms and molecular structure whence the elementary functions are derived. The regular functioning which preserves the individual and perpetuates the species consists of a series of orderly acts in whose fulfilment every cell plays its special part. Each lives, however, more or less independently according to its structure, assimilating nutrition and embodying it in the mould of its own specific constitution.

We do not know the method of molecular differentiation and organization for each species of cell. All we know is that the parts which govern and specialize the working of the cells are the nucleus and the protoplasm, each made up of specific albuminoid substances.

The nucleus rules the activity of the protoplasm and its chief importance is the maintenance of morphological type and the development of the cell. Protoplasm works and assimilates its surrounding medium thanks above all to its plastids or specific granulations, which are differentiated organs proper to each species and frequently varying in nature even within the same cell.

Animal cells cannot from their elements synthetize the albuminoid ingredients of their protoplasm. The proteins arriving within them, or perhaps their more immediate derivatives, are converted into particular alburninoids due to disintegrations, syntheses, and isomerizations which modify, rejoin and assimilate simpler materials absorbed in the digestive tube or arriving from other cells.

In the living cell-protoplasm fundamental protein substances are disassimilated mainly by hydrolysis in a reducing medium and sheltered from any influence of oxygen. From this arises the formation of fresh nitrogenous substances, amides or complex amines, which are definitely converted into urea, carbohydrates and fatty bodies by the effect of simplifying anaerobic reactions.Urea, the definite form in which some fourteen-fifteenths of the total nitrogen in the tissues is eliminated, seems to be produced in the main independently from the oxidation processes.

Between this substance and the protoplasmic albuminoids we have a whole range of intermediate nitrogen compounds. The most complex are proteins still full of considerable activity, such as peptones, diastases and toxins. Afterwards we get creatinic substances and ureides which immediately precede the formation of urea itself.

Furthermore, there are xanthine compounds and uric acid arising from nuclear materials. Many of these derivates are eliminated through the urine or bile whilst others serve for transient syntheses or serve to stimulate digestion or even nervous activity.The ternary products, formed by disassimilation of the protein substances in correlation with urea and other nitrogenous bodies, vanish through oxidation controlled by the oxidases.

The sugars are consumed or changed into fats by loss of carbon dioxide. The fatty bodies are saponified and then oxidized by degrees. It is precisely these oxidations of sugars, fats and other ternary compounds which are the main source of energy for the warming-up and work of the organism.The chemical phenomena, developed in the cell, supply the energy necessary for every manifestation of life. The organization determines merely the direction which this energy must take and the order of the phenomena which preserve the cell, the individual and the species.Reproduced by permission of the publisher

OTTO MEYERHOF

The application of purely chemical methods to the study of the mechanism of muscular contraction has given us the following results:

- Oxidization of the foodstuffs is not the process directly supplying work, but serves solely for the purpose of replacement. Contraction may take place even in the absence of oxygen, at the expense of the energy liberated by anaerobe break-down processes, in which case the decomposition of glycogen makes way for the formation of lactic acid.
- Nor is lactic acid formation the process directly liberating the energy required for contraction; since, as Embden demonstrated, such formation takes place, to a large degree, when contraction has already ceased, and Lundsgaard's experiments showed that through intoxication of the muscle by monolodoacetic acid, the formation of lactic acid comes to a full stop, and the muscle is then still able to carry out a certain amount of anaerobic work. Under such conditions, creatine phosphoric acid, discovered by Eggleton and Fiske, decomposes more intensely than usual and it is this decomposition which supplies the energy necessary for contraction. Whilst in a normal muscle, the

formation of lactic acid starting from glycogen, serves partly to resynthetize the creatine phosphoric acid broken down, just as in respiration, oxidization is employed for resynthetizing the lactic acid to glycogen.

- It is quite likely that neither does the decomposition of the creatine phosphoric acid directly supply the energy required for contraction. At least that is what Lohmann's latest experiments lead us to think. Here again, it may be a matter of a process of recovery whilst some other chemical conversion — decomposition of adenylpyrophosphoric acid into adenylic acid and phosphoric acid — is taking place at the same time as the contraction and appears to be reversed during relaxation.

The idea that the breakdown of adenylpyrophosphoric acid might take place prior to that of the phosphocreatine is due to Lohmann's observations verified not only in intact muscles but also in muscular extracts.

We have already been aware for some years that coupled chemical reactions take place in such extracts and must be borne in mind when changes in energy in the muscle are under consideration.

The following facts can be established in connection with healthy muscle and its histological structure. The break-down of creatine phosphoric acid is impossible unless adenylic acid and pyrophosphoric acid is simultaneously synthetized to adenylpyrophosphoric acid. It is therefore necessary that the first reaction to take place after excitation should be the break-down of adenylpyrophosphoric acid into adenylic and phosphoric acids.

Only then will it be possible for the phosphocreatine to split up so that the adenylpyrophosphoric acid previously decomposed now becomes resynthetized. If the decomposition of the adenylpyrophosphoric acid takes place in between the instant of excitation and the breakdown of the creatine phosphoric acid, we may assume, in agreement with Lohmann, that such reaction takes place simultaneously with the mechanical process of contraction.Reproduced by permission of the publisher

ARCHIBALD VIVIAN HILL

A lecture delivered during the Course in Biochemistry and Physiology of Muscular Contraction and Fermentation at the University of Barcelona (Curs de Bioquómica i de Fisiologia de la Fermentació i la Contracció Muscular — Institut d'Estudis Catalans), 1934.Muscular contraction can be analysed into the four following stages:

First stage: In which the response is produced

Second stage: Lasting throughout the whole period during which the stimulus remains active

Third stage: When the stimulus ceases to act and the muscle relaxes

Final stage: When, under the influence of oxygen, the muscle recovers from the changes taking place during the period of activity

An emission of heat is observable during the rising mechanical response, and a continued production of heat, becoming constant in rate during the period in which contraction is maintained. Coincident with relaxation, the production of heat first diminishes, afterwards again increasing, and finally decreasing to zero as relaxation terminates.

The heat of relaxation represents the disappearance of the potential mechanical energy necessarily present in the muscle in isometric. contraction. One day we shall learn what the chemical reactions are which are associated with each one of these stages.

The anaerobic heat curves can be considered as the energy outline of chemical reactions taking place in the first half minute after contraction, as discovered by Embden, Meyerhof and other investigators.

An interesting fact, not yet explained, is that the maximum intensity of the delayed heat comes first and is more intense after a long stimulation than after a short one. Such a difference in the form of the delayed anaerobic heat appears with such regularity and so plainly that there must be some reason for its occurrence. The amount of this heat is not the same in subsequent contractions: as the muscle gets more and more tired the delayed heat diminishes relatively more than the initial heat which it follows.

This delayed heat is liberated also in the presence of oxygen, but to it is added the heat of recovery which has its source in oxidation and is of longer duration. The ratio between the total heat (produced during and after contraction by a muscle to which sufficient oxygen is supplied) and the heat first liberated (that is, during contraction) gives us important information about the chemical mechanism of contraction.

The total heat set free in the quarter-of-an-hour or twenty minutes during and after contraction maintains a ratio to the initial heat of contraction which was originally thought to be absolutely constant.

This is not exactly so. In the case of a relatively long tetanus the ratio is about 2.4; in a short tetanus, or in a series of twitches, the figure falls to about 2.0; and in an isolated twitch, the ratio lies between 1.6 and 1.7.

From such facts, it may perhaps be deduced that recovery is not invariable, but depends on the nature of preceding activity. The efficiency of recovery is greater in the case of a simple isolated twitch than in that of a tetanic contraction. Perhaps, some day, Professor Meyerhof, whom I see here listening to me, and who has contributed so much to the advance of our knowledge of the chemistry of muscular contraction, will be able to tell us why things happen this way!

A curious phenomenon, discovered by my co-worker, Dr. Feng, is that the metabolism of the muscle at rest increases as the load increases. Taking the load away causes the resting metabolism to drop back immediately. The same effect can be seen on measuring the consumption of oxygen by the muscle.

PLANT GROWTH HORMONES

One of the most striking features of the auxins is the

multiplicity of their actions. Although they were discovered through their growth promoting effect on the Avena coleoptile, especially when applied to one side, and have been assayed in this manner ever since, it was soon made clear that they promote straight growth in many elongating stems and organs, i.e. enlargement along the longitudinal axis. Also, when such organs are slit and placed in a solution of an auxin they undergo differential growth due to greater elongation of the outer than the inner tissues.

The auxins cause formation of roots on stems, they cause cell division in the cambium and the formation of callus at cut surfaces; on the other hand they inhibit the growth of buds, and they also inhibit both the elongation of roots and the formation of the abscission layer on leafstalks and fruitstalks. Further, they trigger in some way the swelling of the ovary into a fruit, as pollination does, so that by applying auxin in the absence of pollination a normal but seedless fruit is formed. Finally, in pineapple but not (so far as known) in other plants, they even cause flowering.

This multiple action is in strong contrast with that of some of the animal hormones which typically cause specific effects on specifically reactive tissues. The multiple action has led to the view, expressed especially by Went and more recently by Gautheret, that auxin acts as a mobilizing agent, bringing specific hormones to the tissue at which it is applied. According to this view, root formation when auxin is applied to the base of a cutting would be due to the mobilization there of a specific root-forming hormone. Experiments of Went, of Bouillenne, and of Cooper have given indirect support to this view, but the support is not strong and there is evidence against it.

For instance, the rooting of small cubes of potato tubers when given high auxin concentrations, or the rooting of fragments of *Helianthus* hypocotyls in tissue culture containing auxin, are difficult to explain on this basis. The influence of leaves in promoting the rooting of cuttings after auxin treatment, brought out especially clearly by Rappaport, was at first thought to be due to the contribution of the specific hormone by the leaves, but in the light of several more recent studies this effect is almost certainly nutritional. In at least one plant which is very hard to root with auxin treatment, the White Hibiscus studied by van Overbeek and coworkers the entire effect of the leaves could be replaced by nutrients, leaving no evidence for a hormonal factor.

Similarly the inhibition of lateral bud development by auxin applied at the tip was explained by the mobilization of bud growth factors at the point of auxin application, thus starving the buds of their essential factors. But inhibition has been shown to occur when auxin is applied to the lateral buds themselves; and although its mechanism is certainly not clear, it does seem to be essentially a direct effect of auxin on the lateral buds rather than a diversion of other hormones away from these buds.

It is, however, possible in this case that the action is due to the intermediate formation of an inhibiting substance by the auxin. Then again, the inhibition of root growth, a phenomenon exhibited by all auxins and in very low concentrations, is observed on isolated roots and is definitely exerted on the root itself.

All these considerations very strongly support the idea which was tentatively put forward more than ten years ago that auxin controls some fundamental *master reaction* in the cell. From this reaction, growth in size differentation, or even inhibition may follow according to the abilities of the cells concerned and their supply of water and other factors. It remains true that during the occurrence of the actual growth process there is some mobilization of materials, i.e. of water and nutrients. But this is a part of the growth process itself, rather than its primary cause. Its part will be referred to later.

Before dealing with this fundamental master reaction and the evidence for it, it will be worthwhile to discuss one of the fields which has been most active in the last few years, namely the relation between *free and bound auxin.*

In many standard test objects growth is a simple function of the auxin applied, otherwise bio-assay would not be possible. From this it is but a short step to the concept that in the intact plant, or rather in a *part* of an intact plant, growth is a simple function of the auxin made available to that part. Of course nutrient materials have to be available too. This simple view had good success in explaining the facts so long as the auxin was determined by diffusion out into agar.

The two-factor scheme given by Went and Thimann for many seedlings, and. founded on diffusion experiments, worked very well. Perhaps the best case is furnished by Dolk's well-known study of geotropism in the coleoptile, where the excess growth of the lower side, which causes the geotropic curvature, was accounted for by the demonstration that 65% of the auxin (instead of 50% when vertical) went to the lower side. Many other organs have given essentially the same results.

The introduction of a new method, i.e. extraction of the auxin from the plant tissue with an organic solvent, has gradually undermined all this. The very first such experiments, on *Avena,* showed auxin all through the plant, including the parts which are no longer growing. This destroys the simple correlation between auxin and growth and

All samples extracted with ether 6 weeks at 2° after the treatment and subsequent acidification. Figures are units of auxin per mg. dry weight.

Table. Auxin in Spinach Leaf Fractions

Fraction	*Treatment*				
	None	*Incubated 24 hours at 37°: at ph 8*	*at ph 10.5*	*at ph 10 With chymotrypsin*	*Boiled at ph 10*
Chloroplasts	0.21	3.54	0.07	9.4	0.16
Globulin i plus ii	0.54	2.00	0.22	21.0	0.26
Albumin	–	2.43	0.80	1.1	0.50

Introduces a distinction between free auxin, which diffuses readily into agar, and bound auxin, which does not. As Went has shown, it is only the former which is redistributed under the influence of gravity. Extension of the extraction method to other material has raised many problems. Let us consider first the case of green leafy tissue. In *Lemna,* for instance, auxin is set free into ether continuously but very slowly. From dried material no liberation occurs, but on wetting it re-starts.

The auxin is apparently being set free by hydrolysis. Proteolytic enzymes, especially chymotrypsin, liberate the auxin rapidly, so that we have here an auxin-protein. Using chymotrypsin we have attempted to locate this auxin-protein in the leaves of spinach, but there is more than one location. Some two-thirds of the auxin is associated with a globulin in the cytoplasm, but there is also much auxin in thoroughly washed intact chloroplasts.

The other protein preparations do not contain large amounts. Bonner and Wildman, using hydrolysis by alkali, found all the auxin in a globulin, a very interesting compound comprising three-quarters of the cytoplasm protein and having properties, including phosphatase activity, similar to those of myosin. Alkaline hydrolysis does not liberate the chloroplast auxin, which explains our inability to extract auxin from *Lemna* by alkali.

Avery and coworkers have reported the extraction of auxin from leaves of cabbage by boiling in alkali; but this cannot be a general phenomenon, for Miss Mes, with tomato leaves, finds as we do with spinach and Lemna that there is no liberation of auxin by alkali.

Very little auxin is liberated by simple short extraction of leaves with ether, i.e. very little is free in the leaf. If this relatively large store of auxin in protein form were available to the plant for growth, it would be expected that leaves would continue growing much longer. One might expect giant cells and other evidences of high auxin concentration; but, further, one would expect to find the auxin very readily set free under natural conditions. Nevertheless many extensive experiments which we performed with different kinds of autolysis or action of autogenous enzymes succeeded only in showing a very small liberation; no large fraction of the auxin was ever obtained even after many days' autolysis.

A measure of the free auxin in the tissue is given by the results of short-time extraction with ether, or in some cases, extraction of boiled tissue with ether. Van Overbeek*et al.* have given evidence that the free auxin is essentially extracted in two hours. From our own data, based on twenty-four hours' extraction, the following figures may be given for the percentage of the total auxin given up in this way, i.e. the presumptive "free" auxin:

Fronds of *Lemna*	2%
The alga *Ulva*	2%
Avena roots	25%
Bean nodules	87%
Avena coleoptiles	Almost 100%

The contrast between the two extremes is very striking, for the auxin in nodules and coleoptiles is very largely in the free state while that in the green tissue is overwhelmingly bound. It is difficult to resist the conclusion that this protein-bound auxin of leaves is *not* a reserve and is (mainly at least) unavailable to the plant for growth.

Seeds, however, show a very different picture. When dry seeds of *Avena* or of corn are dampened, auxin is very rapidly set free. Heating with alkali liberates auxin very rapidly in this material, as shown by Avery and coworkers. It is now clear that this auxin, too, is protein-bound, and Gordon has prepared pure proteins from wheat which liberate auxin on hydrolysis with alkali or acid. We know that seeds deliver to the seedling a precursor which the seedling tip converts to auxin. While this mobile precursor, partially purified by Berger and Avery, does not itself seem to be a protein, it is clear that the auxin-protein of seeds is a true reserve or auxin-source, available to the plant for growth.

A word of warning is in place here. Tryptophane on heating with alkali forms a small amount of indoleacetic acid; casein and other proteins do the same. This phenomenon probably cannot account quantitatively for the auxin liberated by alkali from these seed proteins, but it certainly enters in. The fact that some seedlings are able to convert both tryptophane and tryptamine to free auxin should be remembered also. The demonstration by Haagen Smit, and later by Avery et al., and by Larsen and van Overbeek, that indoleacetic acid is no "heteroauxin" but a true plant auxin is of importance here.

Besides the above instances, there is still another type of behaviour. When grass-stalks are laid horizontal, the nodes, which have long ceased growing, begin to grow on the lower side and this produces geotropic curvature of the stalks. Many years ago Schmitz showed by diffusion that this is due to the new production of auxin in these nodes; here gravity, instead of merely redistributing the auxin, as in the coleoptile, causes renewed *production* of auxin. Now van Overbeek*et al.* have shown in sugar cane that this new production is at the expense of bound auxin in the node; gravity in some way initiates the conversion of the bound to the free form.

This phenomenon of gravitational release of auxin may be connected with another curious phenomenon. In pineapple (but not in other plants) application of auxin to the growing tip causes formation of flowers. Van Overbeek and his coworkers (unpublished) have now found that placing the pineapple plant horizontal has the same effect. This remarkable result may be due to liberation of bound auxin, but could of course be merely the result of geotropic accumulation of enough extra auxin on one side to start the flowering process.

In summary we may conclude that growth depends not only on the free auxin present but also on the ability of the tissue to convert bound auxin into the free form. The bound auxin is itself of many types:

- That of the seed and perhaps the ovary which is made available for growth,
- That of the leaf which is not,
- That of the main axis, which may occupy an intermediate position,
- That in the coleoptile appears much more loosely bound than the types discussed above.

From studies of the structure of the many substances active as auxins, Koepfli, Went have earlier deduced that there is a certain spatial relationship between the various parts of the molecule which is essential for activity. This suggests that the auxin has to combine with another molecule whose spatial structure is fixed, such as a protein, for example. This leads naturally to the idea that the auxin is a coenzyme for some enzyme system which controls growth.

It has been known for a long time that growth of the coleoptile will not take place anaerobically, and Bonner showed that growth is inhibited by cyanide to the same degree as respiration is. This suggests that growth is in some way linked to respiration, but it does not provide a mechanism for studying it. However, Commoner subsequently found that various known inhibitors of dehydrogenase systems also inhibit growth. This led to a study of the whole relationship between enzyme inhibitors and growth.

The technique was the straight growth of isolated coleoptile sections, which grow little in water or sugar alone, more in auxin, and considerably more in the combination of auxin and sugar. Growth was in shallow layers of solution by placing the sections on the teeth of combs.

Of all the inhibitors studied, iodoacetate was one of the most effective, giving complete inhibition at about 3.10^{-5} molal, an exceedingly low concentration for most enzyme inhibitions. Under these conditions the respiration of the sections was very little reduced. We have here not an effect on the general vitality or integrity of the cells, but specifically on the growth process itself. Furthermore this inhibition of growth is removed by malate and other four-carbon acids, and also by pyruvate and (as we have found later) isocitrate.

Also, the growth of the uninhibited sections in auxin and sugar is clearly promoted by four-carbon acids after some hours. It is clear then that growth is not a function of respiration as a whole but of some specific part of it, i.e. of a respiratory system or at least a hydrogen-transferring system, involving those organic acids which participate in the Krebs cycle.Confirmation of this came from the study of protoplasmic streaming in the coleoptile. At first by direct observation with a stop watch and later with a special device which provides an artificial comparison stream whose rate can be adjusted to synchronize with that of the protoplasm particles. The rate of streaming is accelerated by auxin. The reaction is closely related to growth because:

- It takes place before visible growth occurs, being clear within a few minutes,
- Oxygen is required, as in growth,
- Sugar is needed for the effect to last, just as in growth,
- The dependence on auxin concentration is parallel to that of growth,
- The increased rate is prevented by iodoacetate,
- The auxin effect is promoted by malate, especially in old or starved coleoptiles, whose sensitivity to auxin has been greatly reduced.

The parallel is thus very complete; the effect of auxin on streaming is almost undoubtedly a part of its effect on growth. If acceleration of growth and acceleration of streaming are due to a dehydrogenation process, the effect of auxin on respiration should be interesting.

All investigators are agreed that auxin does not raise the oxygen consumption of coleoptile tissue in buffer solution; but we found that in sugar (sucrose and fructose) after a few hours' soaking, auxin does produce a rise of some 10% in oxygen consumption. Furthermore, if the tissue is soaked in malate or fumarate, the rise is larger (about 22%). This effect of auxin on respiration in the presence of malate closely parallels the effect on growth.

The effect of auxin on respiration in presence of sugar has recently been confirmed by Avery and Berger, though not however the influence of malate which evidently needs further study to elucidate the conditions under which it occurs.

The inhibition of growth by iodoacetate and its promotion by fourcarbon acids has been confirmed also for the growth of intact seedlings by Albaum and coworkers. They found certain differences in the time relations, however, which may be due to the use of whole seedlings with accompanying complications of transport, inactivation in the plant, and varying uptake by the roots, or may be due as they suggest to changes in the enzyme systems with age of the seedlings.

Recently Bonner and Wildman have reported that fluoride

inhibits growth, and like iodoacetate the inhibition can be exerted without any marked reduction in total respiration. They were led to this by their finding (mentioned above) that the auxin-protein in the cytoplasm of leaves contains a phosphatase and no other enzyme—a very suggestive fact.

Fluoride and iodoacetate both act on the phosphorylating breakdown of glucose, in fact on directly subsequent reactions, though it is true that in the extracted systems so far studied they act only in higher concentrations than are effective on plant growth. We know from the work of James and his coworkers at Oxford that sugar breakdown in leaves follows closely (though not identically) the chain of reactions worked out in yeast and elsewhere.

It is tempting therefore to postulate that both interfere with growth through the sugar-phosphorylating chain, and this would then be the system that delivers energy for growth. The four-carbon acids might participate by allowing a phosphorylation to occur at the expense of oxidizing malate, etc., as described for succinic, pyruvic, and ketoglutaric acids by Colowick*et al.*and by Ochoa.

In attempting to get some evidence on this we have carefully compared the action of these two poisons on growth. Some time ago it is noticed that iodoacetate, although a growth inhibitor, increases the auxin curvatures in the pea test, although it inhibits straight growth of the same pea seedling internodes at the same cencentration. The curvature may be increased by 50% while growth is decreased by some 50%. This characteristic is shown only very slightly, if at all, by fluoride. The highest nontoxic concentrations of both give opposite effects.

It is found that the inhibition by fluoride is not released by the organic acids, though it can be released by certain hexose phosphates. Our tentative conclusion is that iodoacetate and phosphate inhibit growth in different ways or through different systems. The one affected by iodoacetate is presumably of sulfhydryl nature, and there is other evidence to support this.At this point it is useful to digress for a moment to consider other organisms. If auxins influence growth through such fundamental systems as the Krebs cycle and the phosphorylation of hexose, we should not expect their action to be limited to higher plants. It is important in this connection that the long disputed question whether they control the growth of algae has recently been answered.

Algeus has shown that indoleacetic acid does in fact promote cell division in green unicellular forms and has given an explanation of why up to now the results have been conflicting. One major factor is the strong influence of *p*H on entry of auxin acids into the cell, and the other is that the light necessary for continued growth of the algal cultures produces decomposition products of the auxin which are toxic. If these points are recognized and controlled, clear-cut growth promotion is found. Also Dr. Dubos has informed me that there are in some bacteria very marked effects of auxin on growth. This, together with some new evidence (unpublished) about an influence of auxin on muscle tissue in animals, justifies the view that the action is exerted via systems which are of quite general occurrence.Lastly we must ask how this purely respiratory process can bring about growth.

In the last analysis growth in plants means the intake of water. One way in which this may be brought about is through an effect on the accumulation of solutes. We know from the work of the California school that solute accumulation requires oxygen and consumes sugar and that active protoplasmic streaming is characteristic of cells which can accumulate. Following the demonstration by Reinders that auxin promotes water uptake by potato slices, Commoner and coworkers have shown that auxin even causes the uptake of water from a solution which is hyper-tonic, i.e. a solution which without auxin would cause wilting.

This intake of water closely parallels other effects of auxin on growth, and it is promoted by potassium ions in much the same way as we earlier found the growth of coleoptiles in ordinary auxin solutions to be increased by low concentrations of KCI.Very tentatively, therefore, one may propose the following steps for the simplest kind of growth of isolated plant parts immersed in solutions:

- Auxin enters the cells with great rapidity, accelerating the rate of protoplasmic streaming.
- This acceleration causes increased intake of solutes.
- The energy for the accumulation of solutes and for the streaming is furnished from a sugar breakdown which involves phosphorylation and in which the Krebs cycle participates; one of the key enzymes here is of* the sulfhydryl type.
- The intake of solutes is at once accompanied by intake of water; growing cells do not have any higher osmotic pressure than non-growing ones.
- The acceleration of streaming causes also a difference in the rate of wall deposition, i.e. a change in plasticity.
- The increased water intake coupled with the change in wall properties means increased cell size—in other words, *growth*.

In this way one can begin to see how auxin can have the multiple effects mentioned at the outset. For a cell may not be in a condition to enlarge directly, due either to secondary wall formation or to its unfavorable situation in other tissues, yet the increased streaming rate and increased uptake of solutes may in some way activate numerous processes, causing cell division and leading to root initials and the like. Such a tissue would accumulate organic solutes too, thus providing some basis for the mobilization phenomena which, as was stated above, are probably secondary and not causal.This scheme is still largely hypothetical but there is enough circumstantial evidence to make one believe that it is at least somewhere near the truth.

HUMAN GROWTH HORMONES

The metabolic effects of growth hormone in muscle have been studied in normal and hypophysectomized rats, and, to a large extent in recent years, in the isolated diaphragm muscle. The effects observed have mainly to do with the carbohydrate metabolism of muscle, and these are not as yet clearly or immediately related to the growthpromoting actions of the hormone. The current picture of the effects of growth hormone on muscle is confused by the fact that, depending on the conditions of the experiments, the hormone sometimes seems to promote and sometimes to inhibit, the utilization of glucose by muscle.

The inhibitory effects on occasion seem to be directly mediated by the hormone, but in certain other experiments the hormone appears to give rise to inhibitory substances which circulate in the plasma, so that the effect is indirect and takes time to develop. The details of much of this work have been well reviewed by several workers.

CARBOHYDRATE METABOLISM IN VIVO

One of the striking effects of hypophysectomy in the rat is that the glycogen of voluntary muscle disappears much more rapidly than normal during a fast, falling in 16 to 24 hours to about half the level of the fed state, which is usually not very different from that seen in normal fed rats. The cardiac glycogen, which rises during a fast in the normal rat, also falls to low levels in the fasting hypophysectomized rat.

Since there is rapid depletion of the liver glycogen in the fasting hypophysectomized rat, it might be thought that the failure to maintain the muscle glycogen could be attributed to the early exhaustion of the hepatic sugar store and to the diminution of gluconeogenesis in consequence of the adrenal atrophy after hypophysectomy. In fasting adrenalectomized rats, however, the muscle glycogen stores are well maintained and the cardiac glycogen increases normally, despite the rapid loss of liver glycogen and the diminution of gluconeogenesis characteristic of these animals.

The respiratory quotient of the hypophysectomized rat remains elevated during a fast. If the disposition of a carbohydrate meal is studied in such animals, it is found 4 hours after the meal, that there is less liver and muscle glycogen and a lower blood sugar, and that a much greater proportion of the glucose absorbed appears to be oxidized. Hypophysectomized rats with the basal metabolic rate restored to normal levels with thyroxine exhibit the same abnormalities, but the absolute rate of utilization of carbohydrate is still greater. In the eviscerated, functionally hepatectomized, hypophysectomized rat, the blood glucose and muscle glycogen fall more rapidly than normal.

The rate at which glucose must be infused to maintain the blood sugar constant is about 25-27 mg. per 100 g. per hour as compared to 11-13 mg. per 100 g. per hour in the eviscerated normal rat. Similar rapid rates of glucose utilization were observed in hypophysectomized eviscerated rabbits by Greeley. When it is recalled that the hypophysectomized animal, with its low level of thyroid activity, has a much lower metabolic rate than the normal animal, these differences are all the more remarkable.

If fasting hypophysectomized rats are treated with alkaline extracts of anterior pituitary tissue (APE) or with purified growth hormone during the period of fasting, the fall in muscle glycogen is prevented and the cardiac glycogen rises rather than falls. In fasting normal rats, the injection of growth hormone during the last 6 hours of a 24-hour fast also leads to a greater accumulation of glycogen in the gastrocnemius muscle, the diaphragm, and the heart. In these experiments, there were no significant changes in liver glycogen, so that the effects of growth hormone appear to be exerted chiefly in conserving muscle glycogen. Of the different muscles observed, the heart muscle appeared to be most sensitive to growth hormone, the diaphragm least sensitive.

The effect of growth hormone in conserving muscle glycogen in the fasting animal may be an example of a general tendency of the hormone to diminish carbohydrate utilization. Another aspect of this tendency may be seen from its effects on the disposition of fed carbohydrate in the intact rat. Fasted rats given a carbohydrate meal respond, during the 4 hours after feeding, with a rise in R.Q., and an increase in liver and muscle glycogen and blood sugar.

If insulin is given with the meal, the rise in R.Q. is greater, there is a greater increase in muscle glycogen, and the liver glycogen and blood sugar are diminished. If APE or purified growth hormone is given prior to the meal, there is a still greater increase in muscle glycogen, the blood sugar and liver glycogen are about the same as in the normal rat, but the R.Q. does not rise nearly so high as after carbohydrate alone.

Finally, if both insulin and growth hormone are given with the meal, there is a spectacular accumulation of muscle glycogen, the liver glycogen and blood sugar are intermediate between those seen in the untreated and the insulintreated rats. In experiments in another design, Illingworth and Russell also showed that insulin and growth hormone together had greater effects than either hormone alone on the glycogen of gastrocnemius, heart, and diaphragm of fasted rats given a glucose meal.

All of these observations indicate that growth hormone affects the disposal of available sugar in favor of storage in the muscle rather than oxidation or conversion. It is interesting to note that, although the two hormones have opposing effects on the blood glucose, insulin and growth hormone work together in promoting the storage of glycogen in muscle.

Other aspects of the effect of growth hormone on carbohydrate metabolism are presented by the work of de Bodo and his colleagus on the hypophysectomized dog. The hypophysectomized dog in general behaves in a way quite similar to the hypophysectomized rat; the animals are highly sensitive to insulin, and, in the postabsorptive state, have much lower blood sugars than normal dogs. In experiments using C^{14}-glucose as a tracer, the interesting observation is made that in hypophysectomized dogs 16-18 hours after the last meal, the rate of uptake of glucose by the peripheral tissues and the rate of hepatic output of sugar are lower than normal.

The rate of the CO_2 production in these animals is also much less than normal, as one might expect in hypophysectomized animals with much diminished thyroid function; one wonders, therefore, what portion of the total metabolism is contributed by the blood glucose, and what might be seen if these observations were made with hypophysectomized dogs brought to a normal metabolic rate with thyroxine.

As they stand, however, these experiments give no

evidence of an excessive continuing utilization of plasma glucose in the postabsorptive period. The injection of a small dose of insulin is followed by a greater than normal increase in the rate of uptake of plasma glucose by the peripheral tissues. The ensuing hypoglycemia is relatively even more severe, however, because the liver (which still contains adequate glycogen) fails to respond adequately to the falling blood sugar level.

The exaggerated insulin sensitivity of these animals therefore has both a peripheral and a hepatic component. If hypophysectomized dogs are treated with growth hormone for 4-5 days, the level of the postabsorptive blood sugar rises, the rates of glucose uptake and inflow are restored to normal, the increase in glucose uptake induced by the same small dose of insulin used before treatment is reduced to normal limits, and this is now smoothly compensated by an adequate increase in hepatic output of glucose, an effect evidently brought about by growth hormone but as yet without explanation.

In normal and adrenalectomized dogs also, treatment with growth hormone increases the level of the postabsorptive blood sugar and the rates of inflow and outflow of plasma glucose. In the conditions of these experiments, therefore, although growth hormone reduces the sensitivity to insulin, there is nevertheless an increased rate of flow of glucose from the liver to the peripheral tissues. An antagonism of growth hormone to insulin in the sense of a marked depression of glucose uptake by the peripheral tissues is not seen.

One does not of course know what is happening to the glucose after it leaves the plasma. Data on the specific activity of the expired CO_2 would be most interesting and enlightening in this respect. The picture of these events in the growth-hormone-treated dog is not inconsistent with the observations cited above on the glucose-fed rat treated with growth hormone or insulin and growth hormone. The novel feature of these observations is the effect of growth hormone on the hepatic output of glucose, but speculation on this point is outside the range of a chapter on muscle.

EFFECTS ON DIAPHRAGM AND HEART IN VITRO

In recent years, a great many observations have been made on the effect of growth hormone on carbohydrate metabolism in the isolated rat diaphragm. The results of different investigators have varied a great deal in detail, partly no doubt because of differences in technique, sources of tissue, previous history and strain of animal, and differences in the nature of the hormone preparations used. It has generally been observed that diaphragm from hypophysectomized rats takes up glucose from the medium at a higher than normal rate, although this effect seems to be dependent in part on the glucose concentration in the medium; at the higher concentrations it is definite and reproducible.

There is also general agreement, with one doubtful exception, that growth hormone added *in vitro* to diaphragm from normal or hypophysectomized animals does not depress glucose uptake or diminish the effect of added insulin in increasing glucose uptake or glycogen synthesis. Instead, many workers have noted that growth hormone added *in vitro* has itself an "insulin-like" action, increasing the rate of glucose uptake of diaphragm from hypophysectomized animals.

An insulin-like action of growth hormone may also be observed in normal rat diaphragm, but it is dependent on the nature of the incubation medium. Russell reported that growth hormone added *in vitro* to diaphragm from glucose-fed normal rats diminished the rate of fall of glycogen during incubation, but many subsequent attempts to repeat these experiments have so far failed.

The injection of growth hormone into the animal prior to taking the diaphragm is followed by an interesting series of events. If the tissue is taken within 3 hours, only an "insulin-like" effect on glucose uptake is seen, but thereafter, reaching a maximum in 18-24 hours, the glucose uptake and the response to insulin *in vitro* are much depressed. This delayed effect of injected growth hormone is enhanced by the injection of adrenal cortex extract.

The absence of a direct effect of growth hormone *in vitro* and the development of a depression in glucose uptake in diaphragm removed from the animal some time after the injection of the hormone led to the suggestion that an altered form of the hormone, or some product released into the circulation after its injection, might be responsible for the altered behaviour of the diaphragm toward glucose.

In pursuit of this suggestion, Bornstein and Park observed that the serum of alloxan diabetic rats contained a factor that inhibited the glucose uptake of normal diaphragm. The factor was absent from the serum of hypophysectomized alloxan diabetic rats, but it reappeared after the injection of growth hormone and adrenal cortex extract, but not if either hormone preparation was given singly. Their factor appeared to have the properties of a lipoprotein.

Vallence-Owen and Lukens have made a similar set of observations on plasma from normal, depancreatized, hypophysectomized-depancreatized and adrenalectomized-depancreatized cats. The plasma insulin activity of all of the depancreatized animals, measured by the effect on glucose uptake of the rat diaphragm, fell to zero, and, in addition, there appeared in the plasma of the diabetic cats a factor which greatly inhibited the effect of added insulin (1000 microunits).

The factor disappeared after hypophysectomy or adrenalectomy, and it was not restored by growth hormone alone or by adrenal cortex hormone alone. Since the inhibitory effect of the plasma from diabetic cats was not destroyed by freezing and thawing, it appeared doubtful that it could be a lipoprotein. The relation of this factor to that of Bornstein and Park remains to be determined. Another interesting system in which an *in vitro* effect of growth hormone has been observed is the isolated perfused rat heart. This preparation has the advantages that it has a proper circulation to the cells, an active metabolism, and good sensitivity to both insulin and growth hormone. The rate of glucose uptake by the isolated heart, about 15 mg. per gram dry weight of heart per hour, is depressed about 50% by the addition of 0.1 ¼gm. per milliliter of growth hormone to the perfusion medium.

In the presence of 2 milliunits per milliliter of insulin, the rate of glucose uptake is 40 mg. per gram dry weight per hour;

0.1 ¼gm. of growth hormone per milliliter reduces this to 28-30 mg. per gram dry weight per hour. Studies with galactose indicated that growth hormone had the effect of diminishing the galactose space, whereas insulin has the effect of increasing it. An analysis of the time course of galactose penetration showed that growth hormone, in the presence or absence of insulin, approximately halves the numerical value of a derived coefficient which is taken as a measure of the activity of the carrier system responsible for the entry of glucose or galactose into the heart muscle cells. Insulin, on the other hand, in the presence or absence of growth hormone, increases the activity of the carrier system about fourfold. Since the actions of the two hormones are independent in the sense that each exerts its proportionate effect on the carrier system in the presence or absence of the other, growth hormone is not regarded as an "anti-insulin" but as a factor diminishing the scale of activity of the carrier system on which insulin acts.

The authors point out that the observed effects of growth hormone and insulin on galactose penetration into the heart are entirely consistent with the observed relationship between the adenohypophysis and the pancreas: "First, the depressing effect of factor does not involve insulin; it occurs in its absence. Secondly, the effect of insulin is not dependent on antagonizing factor: insulin is more effective in its absence. There is, in fact, an *in vitro* Houssay phenomenon in our system."

Bronk and Fisher use the term "factor" because the growth hormone preparations used had more than one effect. The less pure preparation used had a complex, biphasic dose-response curve, in contrast to the more highly purified preparation derived from it. Both preparations became more effective on storage in dilute solution at -20°C. for several days, so that some labile factor acting in the same direction as insulin may also have been present. "Factor" *may* be growth hormone, but a decision on this point is quite properly reserved.

This interesting work on an active, intact muscle preparation needs confirmation and extension, but it shows great promise of providing a simplified system in which the analysis of hormonal actions may be carried out under more favorable, more nearly physiological conditions than *in vitro* experiments with thin muscle strips can permit.

The picture of the action of growth hormone on carbohydrate metabolism in muscle is as yet by no means clear. There is evidence that it acts to conserve the carbohydrate stores and that it influences the disposition of carbohydrate within the muscle. There are indications that, in the presence of insulin, the entry of sugar into the muscle may be facilitated (the "insulin-like" effect in the diaphragm), or that (as in the heart) it may act directly to limit the activity of the carrier system that is facilitated by insulin.

The hormone also may give rise to substances which can depress the uptake of glucose by the muscle. In the intact glucose-fed rat, the hormone depresses the R.Q. and spares the oxidation or conversion of glucose in the tissues. Whether this is a direct effect, or a consequence of the mobilization of an alternative substrate, fat, cannot yet be decided.

There are many points of conflict apparent in the present observations, and it is not now possible to resolve them in terms of a single principle. It is evident that a great deal more work needs to be done, in a wider variety of conditions, and certainly on a wider variety of muscles, before a coherent picture of the action of growth hormone on muscle can be drawn.

THYROID HORMONE

That the muscles share in the general effects of thyroxine on the metabolic rate is well established. In the surviving tissues of hypo- or hyperthyroid animals, the muscles and heart have exhibited alterations in oxygen uptake of the same order as those seen in the intact animal, the cardiac tissue showing rather greater changes, skeletal and smooth muscle somewhat less. It is probable that the heart undergoes *in vivo* even greater alterations in metabolic rate than it shows *in vitro*; for in addition to the change in its own "basal" rate of heat production, the work required of the heart, in providing for the altered needs of the rest of the body, must be correspondingly affected. There has been no convincing demonstration that the thyroid hormone affects the metabolism or function of muscle in the mature animal in any other way than through its control of the metabolic rate and consequent effects. The mechanism of action of thyroid hormone on the metabolic rate has not been established.

Metabolism of Foodstuffs

The metabolism of carbohydrate and fat in muscle seems to undergo no special alteration with deficiency or excess of thyroxine other than those expected from the relationship of the supplies of foodstuffs to the demands for energy. The rates of catabolism of administered carbohydrates and ketone bodies and of body fat are affected about in proportion to the metabolic rate, whether it is high or low, and the proportionate disposition of fed carbohydrate in muscle is normal in hypothyroidism.

In hyperthyroidism, the glycogen stores of muscle may be depleted, especially in the heart; but the fall in the cardiac glycogen has appeared to be related quite closely to the work of the heart as indicated by the pulse rate, and similar effects on glycogen have been produced by other means of altering the heart rate. It is probable that similar relations of supply and demand obtain in skeletal muscle and liver as well.

Protein Metabolism

The relationship of the thyroid hormone to protein metabolism is biphasic. On the one hand, as mentioned earlier, small amounts of thyroxine are necessary for normal growth and development, so that in this sense the hormone is anabolic. On the other hand, the "basal" rate of protein catabolism is correlated directly with the basal metabolic rate. This has been seen in the nitrogen excretion during fasting or in subjects on nitrogen-free diets (the "endogenous" nitrogen catabolism); in the rate of release of amino acids from peripheral tissues of eviscerated animals; and in the release of protein and amino acids into the medium from isolated diaphragm.

When fasting thyroidectomized rats were given N^{15}-glycine, a larger proportion of the isotope was excreted and this was diluted to a lesser extent by endogenous unlabeled nitrogen. Hence, both the anabolism and catabolism of body protein appeared to have been diminished. Similar observations in patients with myxedema before and after treatment with thyroid hormone also indicated that the rates of synthesis and breakdown of body protein were both affected in parallel. The rate of weight loss in denervated muscle was slower than normal in thyroidectomized animals, but occurred at the normal rate when the animals were given thyroxine in amounts which restored the normal metabolic rate. The effects of thyroid hormone on the net rate of protein catabolism evidently did not require innervation and were not dependent on muscle work.

In both clinical and experimental hyperthyroidism, a negative nitrogen balance is frequently seen; and if the condition is prolonged, a considerable degree of wasting of the muscles and other tissues can occur. Since the negative nitrogen balance can be overcome by the provision of sufficient food, it is probably mainly the result of imbalance between supplies of foodstuffs and the increased metabolic rate.

One may suppose that if anabolism of protein is increased in this state, it is augmented to a lesser degree than is the rate of catabolism. An increase in the requirements for and secretion of adrenocortical hormones, which has been postulated to occur in hyperthyroidism, might be expected to enhance the tendency to lose nitrogen in this condition.

Creatine Metabolism

Because of the weakness of the muscles characteristic of hyperthyroidism, much attention has been devoted to creatine metabolism in relation to the thyroid hormone. The subject has been reviewed extensively by Wang. Creatinuria, diminished creatine tolerance, and slightly lower excretion of creatinine are usual in both clinical and experimental hyperthyroidism.

The total creatine content of the muscles, particularly of the heart, has been reduced in thyrotoxic animals and the phosphocreatine content also is said to be low. In skeletal muscle, the reduction in phosphocreatine was the same as that in total creatine in absolute terms, but larger proportionately; but the changes were rather small in any case. It may be remarked that in nearly all of the work of this type done in animals, the dosages of thyroid hormone used have been massive in the extreme—i.e., just sublethal or sometimes even fatal in amount. The changes so induced in creatine and phosphocreatine and other energy sources may be more expressive of nonspecific damage than they are of thyroid hormone activity on muscle metabolism per se.

From this work, no special alterations in creatine metabolism or in the functional aspects of creatine in muscle seem to have been uncovered. Since creatinuria and loss of creatine from muscle are seen in a variety of conditions in which atrophy or wasting of the musculature is prominent, it is probable that in hyperthyroidism also these features are related mainly to the loss of substance from the muscles. The relative depletion of phosphocreatine and other energy sources might be expected under conditions of relative overwork, as in the thyrotoxic heart.

MUSCLE FUNCTION

Despite the fact that weakness of the voluntary musculature and other functional myopathies are common in hyperthyroidism, remarkably little work has been done on the fundamental nature of the functional effects of excess or deficiency of thyroid hormone in muscle. In a careful study of the Achilles tendon reflex in man, Lambert*et al.* confirmed the common clinical impression that the contraction is slow in most patients with hypothyroidism and tends to be accelerated in hyperthyroidism. There was no change in the interval between stimulation and the onset of contraction; but both contraction and relaxation were affected, the latter to a greater extent than the former. This suggests that the alterations lie within the muscle, rather than in conduction or transmission.

There have been many reports that the efficiency of muscle work is low in hyperthyroidism—that is, that the increase in oxygen uptake is large in relation to the amount of work done during exercise. Schwartz and Lein have studied the characteristics of muscle contraction simultaneously with oxygen consumption in hypothyroid or moderately hyperthyroid rats. When the rate of stimulation was relatively low (40 per second), the steady-state tension was low in hyperthyroid and high in hypothyroid animals, and the ratio of excess oxygen use to the tension was altered inversely.

It was noted, however, that the fusion of responses at this frequency was quite incomplete in the hyperthyroid animals, complete in the hypothyroid, and intermediate in the controls. This is consistent with the apparent changes in contraction time, a muscle displaying rapid contraction and relaxation, as in hyperthyroidism, being expected to have a high fusion frequency. At higher rates of stimulation, in which fusion was complete in all animals, the steady state tension, although somewhat low in the hyperthyroid animals, was not regularly related to metabolic rate, and the ratio of oxygen use to tension did not differ significantly from normal in any group. From this work, it appeared that the thyroid hormone did not affect muscle economy directly, but that the apparent efficiency of the muscle could be altered by changes in contraction time and degree of fusion of responses to stimulation.

Additional factors which may affect muscle function in hyperthyroidism are, perhaps, the wasting of the tissues in the chronic state, and more important, the probable difficulty in assuring adequate oxygenation in the face of the very great demand which may occur in local areas during exercise. The relative importance of all of these factors in the hyperthyroid state remains to be assessed.

In hypothyroidism, muscle function, although slow, does not seem to be much affected otherwise. The myotonia sometimes described could well be the result of slow relaxation and diminished fusion frequency.

PARATHYROID HORMONE

Although it is well established that the disturbances in excitability that are seen in states of hypo- or hyperparathyroidism are referable to the effects of an altered calcium ion concentration on the nervous system or the neuromuscular junction, there is also the possibility, less clearly recognized but supported by a few observations, that the parathyroid hormone may exert an effect on muscle itself. Brown and Imrie, studying the fall in the excretion of urinary phosphate that follows the infusion of creatine into decerebrate cats, observed that the injection of parathyroid hormone, either 18 hours before or immediately before the infusion was started, brought about a greater and more prolonged fall in the excretion of urinary phosphate.

The injection of the hormone alone was followed by an increase in phosphate excretion, as one would expect. There was also no effect of the hormone itself on the concentration of phosphocreatine or any of the other phosphate fractions of muscle, but when both creatine and parathyroid hormone were given, marked temporary increases in muscle phosphocreatine were observed.

These observations were carried further by Imrie and Jenkinson, who observed that in thyroparathyroidectomized cats, decerebrated, or anesthetized with chloralose, the muscle phosphocreatine is lower than normal, falls to lower levels after stimulation (through the ends of the nerves, previously cut), and recovers only very slowly. Treatment with parathyroid hormone (10 units, twice daily for 2 to 4 days) brought about an increase in phosphocreatine and restored to normal the rate of recovery of the phosphocreatine after stimulation. The authors concluded that parathyroid hormone has an influence on the maintenance of phosphocreatine in the muscle, but they made no comment on the possible mechanism of this effect.

More recently, additional indirect evidence has been obtained which suggests that the parathyroid hormone may have an effect on movements of phosphorus in the body other than those which result from the actions of the hormone on the kidney and the bones. Howard*et al.* observed that, if calcium salts are given intravenously to normal individuals (which it is thought would suppress the output of parathyroid hormone) there is a rise in serum P that is not accounted for by the retention of P by the kidney. The suggestion was that there had been a shift of P from the intracellular to the extracellular compartment. To this may be added the observations of Milne, who found that in hypoparathyroid subjects, the extra P excreted by the kidney in response to the injection of parathyroid hormone did not account for the P which disappearred from the extracellular compartment.

There may therefore have been a shift of P into the soft tissues in responses to the hormone. Finally, in the course of attempts to develop an assay for parathyroid hormone in rats, Teppermanet *al.* observed that the direction of change in the serum inorganic P after injection of the hormone depended on the state of the animal and the route of the injection. In fasting rats, no change in serum P occurred after subcutaneous injection, but a prolonged rise in serum P was observed after intraperitoneal injection of the hormone. In the fed animal, however, intraperitoneal injection was without effect, but subcutaneous injection was followed by a fall in serum P, maximal in 2 hours, which was proportional to the dose of hormone injected. All of these observations suggest that the parathyroid hormone may have affects on tissues other than bone and kidney, and that, since these effects seem to the concerned with the transfer of phosphate, the effects on music metabolism and function may be interesting and important.

The probability that much better preparations of the parathyroid hormone may soon be available makes the renewed investigation of this aspect of the function more attractive. It would be interesting, for example, to learn whether the fall in serum inorganic P which may be induced by carbohydrate feeding is in any way influenced by parathyroidectomy or by the parathyroid hormone.

OVARIAN HORMONES

In addition to their effects on uterine growth and on the development of the endometrium, the ovarian hormones (estrogen and progesterone) influence the functional behaviour of the uterine muscle. An excellent account of the actions of the ovarian hormones on uterine muscle is given by Csapo. The following brief summary is taken largely from his paper, which may be consulted for details and for additional references to the literature.

During estrus, the spontaneous uterine contractions tend to be large and infrequent; in the luteal phase of the cycle, when progesterone dominates the uterus, contractions are very frequent but small. The estrogen-dominated uterus is sensitive to oxytocin, but the progesterone-dominated uterus is refractory to the posterior pituitary hormone. Progesterone therefore tends to block the contractile activity of the uterus, inducing a state of relative quietude which is evidently part of the general protective role of the corpus luteum hormone in the maintenance of pregnancy.

The mechanism of this effect is not yet certainly established. It has been shown that the maximal contractile capacity of the myometrium is unaffected by progesterone, so that the blocking action is not exerted upon the contractile system itself, but is more likely associated with the processes of excitation and conduction. A study of the electrolytes of the uterine muscle reveals that the estrogendominated uterus has a high intracellular potassium concentration, and low intracellular sodium, the ratio K/Na being 5.3.

The progesterone-dominated uterus, however, has a low potassium and a high sodium concentration, the ratio K/Na in the muscle being only 2.9. Since the extracellular potassium and sodium is unaltered in the two states, the effect of these changes is to alter the electrochemical gradients between the uterine muscle cells and their surroundings.

The effect of progesterone is therefore to reduce the membrane potential on the one hand, facilitating local activity, but on the other hand to diminish the propagation of the contractions (an effect of the increase in cell sodium) so that massive full-scale contractions are interdicted. The restoration of the table but fully responsive state characteristic of estrus can occur rapidly after the withdrawal of progesterone because of the ease and speed with which the intracellular ion concentrations may be adjusted. The mechanism by which progesterone brings about the changes in distribution of potassium and sodium is as yet unknown.

Chapter 10

Biochemistry of Muscular Action

The energy used by muscle in performance of work and maintenance of tension is ultimately derived from chemical reactions going on within it. The nature of these reactions was the subject of much experimentation and controversy during the latter half of the nineteenth century, but quantitative and reproducible results were not obtained until the classical work of Fletcher and Hopkins. They showed clearly for the first time that fatigue and death rigor are accompanied by lactic acid production.

This great step forward was due to their recognition of the need to reduce to a minimum any stimulation of the muscle during fixation and extraction. Later, Parnas and Wagner showed that the lactic acid is derived from the muscle glycogen. Now it was known that the formation of lactic acid from glycogen is a reaction going on with output of heat—the difference in the heats of combustion of the two substances is 16,300 cal. per gram molecule lactic acid formed, according to Meier and Meyerhof. It seemed reasonable to suppose—and the assumption has proved correct—that here was a reaction providing energy necessary for contraction.

The long series of studies in Meyerhof's laboratory afforded support for this view, showing as they did the proportionality during anaerobic contraction between lactic acid formation and tension production or work done.

The experiments on heat production in frog muscle during anaerobic contraction and relaxation (the initial heat) showed that the heat production was much greater than the expected amount—about 35, 100 cal. per gram of lactic acid formed. A part of this excess could be explained by neutralization in the tissue, but a discrepancy of some 45% remained until it was explained by the metabolism of phosphagen. The significant fact was discovered that the size of the initial heat, its distribution in time, and its relation to tension production are the same when the muscle contracts in oxygen as when it contracts in nitrogen.

Thus the chemical reactions underlying contraction must be nonoxidative; later D. K. Hill showed for frog muscle at O°C. with a short tetanus that even when conditions are from the beginning aerobic, increased oxygen consumption does indeed only start after activity (contraction and relaxation) is over. The aerobic recovery-heat production is large, almost equal to the initial heat; but if the conditions are maintained anaerobic, the period after activity shows only small and variable heat production—averaging about 20% of the initial heat, spread over about 30 min.

CONTRACTION AND CREATINE PHOSPHATE

For twenty years, the idea of lactic acid formation from carbohydrate as the only energy-providing reaction remained unchallenged. Then Lundsgaard found that lactic acid formation in muscle could be stopped by poisoning with iodoacetate, but that nevertheless contraction could go on. He showed that in these poisoned muscles, tension production was proportional to breakdown of creatine phosphate.

This substance (phosphagen) had been discovered in muscle by Eggleton and Eggleton and independently by Fiske and Subbarow who first elucidated its structure. Eggleton and Eggleton connected creatine phosphate metabolism with contraction, for they found it to decrease in amount during contraction, while resynthesis occurred on recovery in oxygen.

Nachmansohn showed that even in nitrogen there was rapid resynthesis of about 30% of the creatine phosphate in the 30 sec. immediately after relaxation. At the same time, Meyerhof and Lohmann found creatine phosphate hydrolysis to be an exothermic reaction, about 12,000 cal. per gram molecule of inorganic phosphate being liberated in the conditions of their experiments. Nachmansohn, comparing tension development with creatine phosphate disappearance, found that during a succession of contractions, this disappearance is much greater during the early contractions than during the later ones.

These facts concerning creatine phosphate necessitated a review of the lactic acid theory of contraction. In the first place, all experiments up to that time had seemed to show proportionality between tension production, heat liberation and lactic acid formation; but if creatine phosphate hydrolysis, an exothermic reaction, was going on to a greater degree at the beginning of a contraction series, a greater heat production at this time was to be expected.

Again, some explanation was needed for the rapid anaerobic resynthesis of creatine phosphate immediately after contraction. Here was an endothermic reaction going on during a short space of time when heat exchange was negligible and certainly no equivalent heat absorption could be detected. The suggestion was made at the time that creatine phosphate was merely "unstabilized" *in vivo*, some change rendering the compound unusually liable to be broken down by the chemical treatment used in estimation.

Then Lipmann and Meyerhof made the significant observation that, during the first few of a series of short tetani, there is a change in the muscle pH not to the acid but towards the alkaline side. These pH changes were studied in intact uninjured muscle (thin frog sartorii) lying in a bath of bicarbonate Ringer solution with a nitrogen- CO_2 atmosphere above. They were related by Lipmann and Meyerhof to an earlier study of the titration curves of creatine phosphate and of an equimolecular mixture of free creatine and inorganic phosphate. Comparison of these curves showed that between pH 3 and 7.5, hydrolysis of creatine phosphate is accompanied by liberation of base; it was striking that the degree of pH change in the muscle on contraction varied according to the pH of the bathing medium and closely paralleled the amount of base liberation to be expected at different pH values from the titration curves. Thus it was shown that creatine phosphate breakdown in the muscle is a reality, and the way was prepared for the results of Lundsgaard.

Lundsgaard took the point of view that, in muscle poisoned with iodoacetate (which prevents lactic acid formation by inhibiting glyceraldehyde phosphate dehydrogenase), creatine phosphate breakdown supplies the energy for contraction; he went on to suggest that this might be the normal rôle of creatine phosphate hydrolysis in unpoisoned muscle also, the rôle of carbohydrate breakdown being to supply energy for resynthesis of creatine phosphate.

In further experiments on contractions of very short duration or of a lightly loaded muscle, he showed that in these circumstances lactic acid formation was less in proportion to tension than under conditions of greater total tension production. These results, taken together with Nachmansohn already mentioned, show that although the heat/tension ratio remains constant, the chemical reactions responsible for the heat production vary with time in a contraction series.

With regard to the anaerobic recovery, Lehnartz brought convincing evidence that much of the lactic acid production takes place after the contraction is over. Such claims made earlier by the Embden school had been disregarded, since with the strength of direct stimuli used a pathological condition of some fibers supervened, with prolonged relaxation time and incomplete recovery. Lehnartz now used stimulation through the nerve and established that 20 to 30% more lactic acid was formed in the 5 min. after relaxation.

Indeed, at low temperatures, more than half the lactic acid formation may take place after relaxation. In this lactic acid production, we have the energy source for the anaerobic resynthesis of creatine phosphate.

It is interesting that several other phosphagens have been found, all guanidino compounds; it is very probable that others still unknown exist. Creatine phosphate is found in all vertebrate muscle examined and also in some invertebrates. Arginine phosphate is the most widely distributed of the rest, being characteristic of most invertebrate muscle. Recently, however, the discovery has been made that certain invertebrates contain, as well as creatine phosphate, new phosphagens; thus in the annelids, glycocyamine phosphate has been identified in *Nereis diversicola* and taurocyamine phosphate in *Arenicola*. Hobson and Rees have shown that in these animals phosphokinases are present which can bring about the phosphorylation of the bases in question by means of ATP. Thoai and Robin have isolated guanidylethylseryl-phosphate from the earthworm; chromatographic examination showed the presence of the corresponding phosphagen in its muscle. From leech muscle, a new hitherto unidentified guanidino compound has been isolated by Robin *et al.* since it is the only guanidino compound present, it is likely that it functions as a phosphagen. In each phosphagen, one hydrogen of the terminal amino group is replaced by the phosphate group.

THE DISCOVERY OF ADENOSINE TRIPHOSPHATE

Soon after the isolation of phosphagen there was discovered in muscle, independently by Lohmann and by Fiske and Subbarow, the substance adenosine triphosphate (ATP). Lohmann studies indicated the structure shown in Table; this has been confirmed by its synthesis. Two observations of importance were made; in the first place, that ATP acted as a coenzyme of glycolysis, though the details of its participation were not worked out till later; in the second place, that hydrolysis of the two terminal phosphate groups led to liberation of heat—about 12,000 cal. per gram molecule of phosphate, according to Meyerhof and Lohmann.

The realization of the importance of ATP hydrolysis for the contraction process itself came only later, and arose out of the work of Lohmann on hydrolysis of creatine phosphate in dialyzed cellfree muscle extracts. Such extracts cannot cause splitting off of phosphate from creatine phosphate; this only happens if adenylic compounds are present and the reaction occurs in two stages.

Creatine phosphate + ADP $\rightarrow$ creatine + ATP (1) ATP $\rightarrow$ ADP + H_3PO_4 (2)

The first is a transfer of phosphate to adenosine

diphosphate (ADP), then hydrolysis of the ATP thus formed goes on to give ADP again. It should be mentioned here in parenthesis that, in all the earlier work, adenosine monophosphate (AMP) was used in reaction systems which, as we now know, require ADP. The adequacy of the AMP is explained by the presence in such systems of the enzyme myokinase which enabled the AMP to react with traces of ATP.

ATP + AMP → 2 ADP

This work of Lohmann had consequences of great significance. Thus he deduced that before creatine phosphate breakdown can yield energy, ATP hydrolysis must have occurred; this latter reaction thus became the energy-yielding reaction closest to contraction. Further, this was the first observation of phosphate transfer, and involved two compounds each containing what we now call an "energy-rich phosphate bond".

Transfer of phosphate between such molecules without formation of inorganic phosphate is a mechanism for conservation of free energy which has turned out to be of enormous biological importance. At the time, Lohmann pointed out that the phosphate transport in reaction (Lohmann's reaction) went on with very little heat exchange; he expressly remained noncommittal about free energy changes.

To provide for resynthesis of ATP, then, was the rôle of creatine phosphate in muscle metabolism. Parnas next initiated an enquiry into the mechanism whereby carbohydrate breakdown provided energy and phosphate for the rephosphorylation, perhaps of creatine, perhaps of adenosine diphosphate.

It was known that hydrolysis of phosphopyruvic acid was an exothermic reaction (about 9,000 cal. being liberated per gram molecule of phosphoric acid released) and this seemed a likely stage; it was in fact found in muscle extracts that phosphopyruvate (like creatine phosphate) transferred phosphate to adenylic compounds and was not dephosphorylated when these were absent.

Phosphopyruvate + ADP → pyruvate + ATP

In the presence of creatine as well as a catalytic amount of ADP, creatine phosphate synthesis took place, but no direct reaction between phosphopyruvate and creatine was found. It follows that reaction must be reversible, and this reversibility has been directly demonstrated. The equilibrium point depends on the pH, more alkaline reactions favoring creatine phosphate synthesis. It seems then that in the recovering muscle, once the stimulation to ATP breakdown has ceased, the phosphate from carbohydrate intermediates is transferred through ATP mediation to free creatine to rebuild the creatine phosphate store.

Now Lundsgaard had found that for every molecule of lactic acid formed in the anaerobic recovery period, about two molecules of creatine phosphate were resynthesized. The reaction we have discussed could account for only half this resynthesis. But there is another exothermic reaction going on in glycolysis, the oxidoreduction between glyceraldehyde phosphate and pyruvate, giving as end products phosphoglyceric and lactic acids. This was known to be accompanied by esterification of inorganic phosphate; it was now found that, if adenylic acid was added to the oxidoreduction system, stoichiometric synthesis of ATP went on, one molecule of phosphate being esterified for every molecule of lactic acid produced.

When this reaction also is taken into account, the extent of creatine phosphate synthesis in anaerobic recovery is readily understood. Although it has been known for some years that 1,3-diphosphoglyceric acid is formed in this coupled esterification and acts as the phosphate donor to ADP, the sequence of reactions has only recently been made clear. It involves reaction of the aldehyde group of the glyceraldehyde phosphate with the SH group of the enzyme, glyceraldehyde phosphate dehydrogenase; the oxidation of this hemimercaptal with formation of an "energy-rich" bond; and then phosphorolysis of the enzyme-acyl compound by means of inorganic phosphate to form diphosphoglyceric acid, which can transfer its acyl phosphate to ADP.

It is of interest to notice that in the case of 1,3-

diphosphoglyceric acid, a direct transfer of the acyl phosphate to free creatine has recently been demonstrated in preparations from rabbit muscle, the intermediation of adenylic compounds being excluded. The phosphokinase involving transfer to ADP was also present and of greater activity. The direct transfer system to creatine was apparently absent from the heart, a tissue depending primarily on aerobic metabolism.

The studies made by Meyerhof and his colleagues on the heats of hydrolysis of compounds involved in muscle metabolism led to the early distinction between compounds containing the guanidino-, pyro-, or enolphosphate linkage on the one hand, and the phosphate ester linkage on the other. Hydrolysis of the last was found to liberate only about 3,000 cal. per gram molecule of phosphate.

It should be noted here that the heat of hydrolysis of ATP has been re-evaluated by several workers during recent years and considerably reduced. The latest value (found by enzymatic attack *in vitro,* in buffers of known heats of ionization) takes into account the heat of neutralization of H^+ ions produced during the reaction and amounts to only 4,700 cal. per gram molecule of H_3PO_4 set free.

Redetermination of the values for the other heats of hydrolysis mentioned above is no doubt necessary; they have been quoted here because of their historical importance in the development of ideas on energy provision. The important aspect of these reactions is, of course the free energy and not the heat change; this was realized at the time, but since methods were not then available for measuring the free energies, the heats of reactions were taken as a rough guide. Since the treatment of this subject by Lipmann, much effort by many workers has been put into the important task of finding true values for the free energy of hydrolysis of these compounds.

We have then this picture of the sequence of events after the stimulus reaches the muscle—first, dephosphorylation of ATP which, though masked by resynthesis in moderate contraction, is usually considered to be the essential reaction. As we shall see, there is good reason for postulating a direct reaction between the ATP and the myofibrillar protein actomyosin. Indeed, the ATP may break down in this initial reaction by a transfer of phosphate to some protein site. This is followed at once by reaction between creatine phosphate and ADP; later, rephosphorylation of ADP by phosphopyruvate and diphosphoglycerate from carbohydrate breakdown becomes quantitatively more important.

Attention has been concentrated here on contraction under anaerobic conditions, depending ultimately on glycolysis; this is because the reactions concerned, unlike many oxidative reactions, can go on readily in cell-free extracts, and so gave the first insight into problems of energy provision. But although anaerobic contraction must sometimes happen *in vivo*, conditions are much more usually aerobic, and oxidative rephosphorylation of ADP is far more efficient.

CONTROL OF THE METABOLIC RATE ON STIMULATION

The presence of ADP may be considered as an adequate trigger, at pH values around 7, for the breakdown of creatine phosphate; its importance in regulating oxidative activity will be considered in Chapter. But it is not so clear what mechanism starts the rapid breakdown of glycogen which begins on stimulation.

The adenine nucleotides are not involved in the earliest stages of glycogenolysis and adequate inorganic phosphate to saturate phosphorylase (the enzyme phosphorylating glycogen) is already present in the resting muscle. C. F. Cori has emphasized the enormous increment in rate of glycogenolysis following stimulation. For example, with frog muscle in nitrogen at 20°C., ten contractions per minute cause a thirtyfold increase over the resting rate and a 10 sec. tetanus causes a thousandfold increase.

Now the enzyme phosphorylase exists in two forms, known as the *a* and *b* forms, the former active and the latter inactive without adenylic acid. An enzyme is present which brings about inactivation of phosphorylase a: the enzyme molecule is split giving a product of half the molecular weight, phosphorylase b; at the same time inorganic phosphate is liberated. Fischer and Krebs showed also the presence in muscle extracts of an enzyme activating phosphorylase b in presence of ATP and Mg^{++} by transfer of phosphate to the enzyme.

Cori believes both enzymes, the phosphorylase-rupturing and the phosphorylase *b* kinase, to be concerned in *vivo* in regulation of carbohydrate metabolism. With stimulation of rat gastrocnemius under controlled conditions, increase in active phosphorylase was seen, while fatigue led to a decrease. During a recovery period of 10 to 20 min., the level of phosphorylase a rose again.

It is not at first apparent why the activation of phosphorylase should lead to increased glycogen breakdown, since an equilibrium reaction is involved:

Glycogen + $H_3PO_4 \rightarrow$ glucose-1-phosphate

In order that there should be increased breakdown, increased rate of removal of hexosephosphate is probably also necessary. Indeed there is some evidence that the phosphohexokinase stage is limiting in resting muscle. Activation of this enzyme may result from the raised Mg/ATP ratio at some locations in the muscle cell as ATP is used in contraction.

INTERACTION OF ADENOSINE TRIPHOSPHATE

During the years that saw this preoccupation with the energy sources of contraction, a parallel and quite independent line of work had been pursued in the study of the muscle proteins. As early as the middle of the nineteenth century, Kühne had extracted from muscle a protein, capable of gel formation, which he called myosin. This protein was soon recognized as belonging to the class of globulins—soluble in NaC1 or KC1 solutions, but precipitated by dilution with large volumes of water.

Edsall and von Muralt and Edsall purified the protein and studied its physicochemical properties, particularly the double refraction of flow of its solutions. Smith investigated the dependence of the solubility of myosin on salt concentration and on pH. He concluded that under the conditions prevailing in resting muscle (pH about 7 and salt concentration equivalent to about 0.18 *M* KC1), at least 90% of the myosin must be in the gel form. About the same time, H. H. Weber was making myosin filaments by squirting the solution in 0.5 *M* KC1 through fine extruders into a large volume of water, so that dilution precipitation took place.

It was not until some years later that Engelhardt and Ljubimova brought the two lines of work together by their discovery that myosin, prepared and purified by the classical methods, is a specific ATPase; when activated by certain divalent ions, it splits off the terminal phosphate group to give ADP. This was very quickly followed by the finding that there is a reciprocal action of the substrate on the enzyme protein. Thus Needham*et al.* observed that addition of ATP to a solution of myosin (made by many hours extraction of the muscle with salt solution) led to a striking fall in viscosity and double refraction of flow—changes which are to be interpreted as a diminution in the axial ratio of anisometric molecules in the solution.

These changes were reversed as the ATP was destroyed by the enzymatic activity of the myosin. The tentative suggestion was made that a shortening of the myosin molecule took place and might be the basis of contraction. But in Szent-Györgyi's laboratory in Hungary, where intensive study of these questions was also going on, it was shown that two proteins are concerned: the protein which we now term myosin, extractable from minced muscle by short treatment (20 min.) with salt solution; and the protein actin, extractable from the dried residue.

Each of these proteins alone can give a solution of low viscosity and without double refraction of flow. On mixture of the two, a solution of high viscosity and strong double refraction of flow was obtained. The conclusion was drawn that the two proteins enter into some form of combination, giving actomyosin, and that the changes seen on adding ATP are due to dissociation of the complex, with formation of the two types of protein molecule of smaller axial ratio.

The relevance of the actomyosin-ATP relationship to contraction was more clearly brought out by the experiments of Szent-Györgyi in which ATP was added to actomyosin at low ionic strengths, in fact to actomyosin gels. Actomyosin threads (prepared by Weber's method) were used and an isodimensional contraction to about 10% of the original length was obtained. If the micelles in the thread are oriented by partial drying and stretching, the effect of added ATP is to make the thread become shorter and wider—as happens in contraction of a muscle fibre.

THE INTERACTION OF ATP AND ACTOMYOSIN IN SOLUTIONS

At the basis of most thinking on the contraction-relaxation mechanism, we find the conception outlined above—that addition of ATP to an actomyosin solution leads to dissociation of the two proteins, while removal of the ATP is followed by recombination. Bailey and Perry have given evidence for the view that the ATP displaces the actin, combining like the actin with sites on the myosin in the neighborhood of essential SH groups. Direct evidence for this dissociation conception was, however, for a long time lacking, but this gap has recently been filled.

Thus A. Weber showed that if actomyosin is centrifuged (under conditions in which ATPase activity is inhibited) with ATP for 3 hours at 100,000 × *g*, pure myosin could be recovered from the upper half of the supernatant and identified by its ATPase characteristics, its reaction with actin, and its sedimentation constant.

The pellet on extraction yielded actin of characteristic behaviour. Gergely has come to the same conclusion from light-scattering experiments. The study of changes in light scattering of a system provides a rapid method of following changes in the size, shape, and interaction of the particles of the light-scattering material. By application of the extrapolation method of Zimm, particle weight can be determined independently of any assumptions as to particle shape. In this way, Gergely obtained evidence of a large fall in molecular weight upon ATP addition to actomyosin solutions, consistent with dissociation. The results with this method are, however, still the subject of controversy.

Mommaerts has shown that the viscosity drop on ATP addition can be obtained without any accompanying enzymatic dephosphorylation; the change is therefore due to combination of enzyme and substrate without breakdown of the substrate. Thus 10-3 *M* Mg^{++}, which inhibits ATPase activity of the actomyosin solution, allows the unimpaired viscosity fall, but there is no recovery.

More recently, he has studied the same change in the protein particles by the more sensitive fight-scattering method and has shown that Ca^{++}, while accelerating the ATPase activity, greatly decreases the light-scattering fall. The high rate of ATP disappearance may have been partly responsible for the diminished fall; but the results of Baranyi*et al.* using a method of very rapid measurement of viscosity changes, show that there really is a very marked activating effect of Mg^{++} ions and a smaller inhibitory effect of Ca^{++} ions on combination between actomyosin and ATP. This point will be raised again later.

Straub and Mommaerts further showed that inorganic pyrophosphate which is not hydrolyzed and inorganic triphosphate which is only very slowly hydrolyzed can in certain circumstances show the unreversed viscosity fall. A high concentration of Mg^{++} (0.01 *M*) is necessary here and Ca^{++} is ineffective.

THE INTERACTION OF ATP AND ACTOMYOSIN IN GELS

THE TYPES OF SYSTEM USED

We have already mentioned the use of the actomyosin thread after orientation by partial drying in a stretched state, as a model of the muscle fibre. An even more useful preparation, introduced by Szent- Györgyi, is the glycerinated fibre bundle. A strip of muscle about 2 mm. in diameter is removed, being kept at the resting length; from the psoas of the rabbit, for example, a bundle of parallel fibers some 8 cm. long can easily be obtained. The fibre bundles are kept in 50% glycerol at 0°C. for some days.

In this way, the removal of water from the muscle is very gradual and the water content remains uniform throughout the bundle; about 50% of the soluble proteins and most of the crystalloids are extracted. An important characteristic is that, owing to destruction of membranes, the fibers show no response to electrical stimulation but are permeable to ATP. Such fibre bundles can be kept for many weeks; they can be washed free from glycerol when needed, dissected to smaller dimensions, and then used for the study of tension production upon addition of ATP. The tension production and degree of shortening are quantitatively very similar to the responses of living muscle to electrical stimulation.

THE STAGES IN THE INTERACTION AND THE DUAL RÔLE OF ATP

The first investigation of the effect of ATP on such muscle models was made by Engelhardt*et al.* on actomyosin threads. These contained only 2% of protein but showed a considerable amount of tensile strength. When a load of some milligrams was applied by means of a torsion balance to the thread immersed in a bath, extensibility could be measured. Addition of ATP to the bath led to an increase in extensibility by some 50 to 100%. At first sight, these results seem to be in contradiction to those of Szent-Györgyi, made a little later, in which as we have seen, the effect of ATP was to cause contraction.

Some years later, Buchthal*et al.* showed that the same threads (in this case 20% protein) would give both effects—if loaded, e.g. with 200 mg., they showed extension on ATP application; but unloaded, they contracted. By very carefully controlled drying and stretching, it is possible to prepare actomyosin threads which will develop considerable tension, but the point made here is that the extensibility increase with ATP indicates that the first effect on the gel is a loosening of linkages, as with the solution.

There is much evidence, as we have seen, that in the case of the solution, myosin and actin are dissociated from one another; while the same degree of dissociation cannot take place in the gel, it is quite possible that the same bonds are affected. Szent-Gyrgyi, using glycerol-extracted muscle fibers, emphasized this dual rôle of ATP. After treatment with ATP the contracted fibre bundle was hard and opaque, but on renewal of the ATP it became momentarily soft and flexible before hardening again as the fresh ATP was used up. Bozler was the first to get a full cycle of contraction and relaxation; here, with the glycerinated fibre bundle, a high ATP concentration was used (0.02 *M*) and the contraction was brief. A concentration of 0.003 *M* ATP caused only contraction.

A better understanding of the dual rôle of ATP in muscle began with the work of Marsh. He used fresh muscle homogenates, in which the fragments consisted of fibre bundles, and he followed their changes in size by centrifuging and measuring the volume of the solid layer. When a brei from fresh, actively glycolyzing muscle was used, the fibre volume remained constant for a considerable period, depending upon the glycogen content of the muscle. After 20 to 50 min., there was a rapid fall in volume, then again a steady state. He considered that the fibre shrinkage corresponded to contraction, the water loss which occurred from the fibers being a consequence and not the cause of the diminution in volume.

Experiments in which the behaviour of the fibre fragments was examined under the microscope bore out this point of view. With fresh homogenates, ATP addition usually led first to an increase in volume and length of the fibers (relaxation); only later, after a time during which diminution of the ATP concentration within the fibers (through their own low ATPase activity and that of soluble ATPase) would have occurred, did the fibers shorten. The rise and fall in volume could be repeated several times.

Marsh was led to emphasize the importance of ATPase activity by the observation that if the fibers were washed twice with salt solution, the only response to ATP added to the suspension was decrease in volume, never increase. At the same time, the ATPase activity of the fibers rose some tenfold as a result of the washing. Marsh deduced the presence in the original brei of a relaxing factor responsible for these effects and we shall return to this later. He also deduced that dissociated actin and myosin in the relaxed muscle must change to actomyosin before contraction; for this change and the shortening, both an increased rate of energy liberation from ATP and a fall of ATP concentration at the active sites on the protein are necessary. He did indeed find that the contraction only occurred when the ATP concentration within the fibers had fallen.

Thus while earlier work on the effects of ATP on gels simply laid emphasis on the shortening response upon application of the nucleotide, later the question of the correlation of the effect with ATPase activity was investigated. We shall discuss this further. The system used by Marsh was too complex to allow proof of his deductions, and there has been much further work on these questions. Meanwhile we may anticipate the point of view which will be developed in what follows by considering in stages the action of ATP on actomyosin gels in vitro, as we have already done for its action on actomyosin solutions.

First we have the loosening of linkages. Second, ATPase activity supplies energy and reduces the ATP concentration at the active sites. Third, the formation of new linkages bringing about the shortening is now possible. From all the evidence of Section I, we must conclude that continuous ATP breakdown is necessary as long as tension production goes on. ATPase activity will continue until all the ATP is used up, and the fibre then remains in the shortened form. Under certain conditions, fresh ATP can cause relaxation.

PROBLEMS OF DIFFUSION

In experiments such as those of Marsh, the fresh fibre fragments can for a time keep up their own internal ATP concentration by carbohydrate metabolism; with experiments with the artificial thread or the glycerinated fibre, there is dependence from the beginning on the balance between the rate of diffusion inwards of the ATP on the one hand, and the rate of hydrolysis by the actomyosin on the other. In the steady state, the concentration of ATP on the outside (denoted by C) is related to the concentration at the centre (denoted by *I*) according to the Meyerhof-Schulz formula:

$$C = Ar^2/4D + I$$

where A = rate of splitting, r = radius of the fibre, D = diffusion constant of ATP within the fibre.

Calculation shows that with a physiological concentration of ATP (about $5 \times 10^{-3}M$), the diameter of a fibre must not exceed about 6 μ if the concentration is to be the same in the centre. With a fibre bundle 500 μ in diameter, the concentration at the centre would be zero. In such a case, the fibers of a large central core remain stiff and play no part in the contraction, indeed even hinder it. Tension measurements have been successfully carried out with thin single fibers, for example by Briggs and Portzehl; here the fibers were 50 to 60 μ in diameter.

They showed a much higher tension production per square centimeter, about twice as great as with fibre bundles, presumably because the inactive core is much smaller. Even so, it was calculated that the ATP-free core would be 20 to 50% of the cross section; but it is much less likely than in a thicker fibre to be in a condition of rigor since it has available ample supply of ADP, a good plasticizer in the presence of enough Mg^{++}. It will be obvious that correlations of tension production and ATPase activity will present difficulties when fibre bundles are used, since the exact conditions of diffusion will vary from one bundle to another.

When on the other hand, comminuted glycerinated fibers or isolated fresh myofibrils some 2 to 3 μ in diameter are used, there is no diffusion problem, and enzymatic activity under different conditions can be accurately studied. Correlation of this ATPase activity with rate or degree of shortening is sometimes usefully made, but it is impossible to assess how far these mechanical effects can be taken as an indication of the power to produce tension.

ATPASE ACTIVITY OF MYOSIN AND ACTOMYOSIN

Purified myosin, free from actin, has ATPase activity, but actin has none. It is therefore assumed that the ATPase activity of actomyosin, though it shows some characteristic differences, is due to sites on the myosin component. An important difference between actomyosin and myosin ATPase is that the former is activated by Mg^{++}, the latter not. This Mg^{++}-activation is seen with actomyosin at low ionic strength; above about 0.2, addition of ATP leads to dissociation so that then the enzymatic characteristics of myosin ATPase appear. Both ATPases are activated by Ca^{++}.

In view of the various indications that Ca^{++} and Mg^{++} ions are concerned with different sites on the myosin molecule, it is of interest to notice that the activation energy of the Mg^{++}-activated ATPase is much greater than that of the Ca^{++}-activated. Bendall (personal communication) has found an increase in rate of some 400-fold with the former on passing from 0°C. to 35°C.; with the latter, the increase was only some 20-fold. Hasselbach and Perry and Chappell have also noticed the high temperature coefficient of Mg^{++}-activated actomyosin.

The ATPase activity of these two proteins has some special features which maybe of importance, as we shall see, in contraction and relaxation.

THE HIGH INITIAL RATE OF A TPASE ACTIVITY

Weber and Hasselbach, using glycerinated fibers 30 μ thick, showed that the ATPase activity was at least twice as great during the first 15 sec. of reaction at room temperature as later—after about 100 sec., when a constant rate was reached. Accumulation of end products, decreasing concentration of ATP, impurity in the ATP, and contraction of the originally relaxed fibers were all considered ruled out as causes of the decrease in rate. The effect could be repeated if the hydrolysis was interrupted by Salyrgan inhibition, then reactivated by cysteine; it was in fact obtained only with fibers newly beginning to cause splitting.

The authors suggest that there may be slow reversible formation of an inactive enzymesubstrate complex of the sort described by Chance with catalase. The experiments described were done with Mg^{++} activation, though a case is given where Ca^{++}-activated myosin gave a similar but smaller effect. Bendall (personal communication) has found the effect to be consistently much smaller with Ca^{++} than with Mg^{++}-activation of myofibrils.

PHOSPHORYLATION OF THE ENZYME

On analogy with its other important biochemical reactions, it has often been suggested that, in ATP breakdown, the first stage is transfer of its terminal phosphate to some site on the myosin—or perhaps on the actin. If this mechanism does indeed operate, one should be able to find exchange between ADP^{32} and ATP in presence of the enzyme. Koshland*et al.* were unable to find such a reaction catalyzed by Ca^{++}-activated myosin or actomyosin.

Thus if a phosphorylated protein intermediate is formed under these conditions, its existence must be transitory. This, as H. H. Weber has emphasized, is only to be expected, since a virtually irreversible, energy-yielding reaction would probably follow at once; moreover, any labeled ATP formed by the back reaction would be in a more favorable position for dephosphorylation than incoming ATP, since the former would be already in the neighborhood of the active sites.

Ulbrecht and Ulbrecht, using highly purified natural actomyosin or isolated myofibrils washed very thoroughly, showed that an exchange could readily be detected provided that Mg^{++} was the activator. However, this exchange was found to be independent of the presence of myosin—it continued in myofibril preparations from which the myosin had been removed. It is difficult to conclude that this means actin phosphorylation, since purified actin showed no exchange, although of course it can combine with myosin to give an actomyosin of normal ATPase activity and normal contractility in the gel form. The situation thus remains obscure.

SUBSTRATE INHIBITION

The existence under certain conditions of an optimal

concentration of ATP for the ATPase activity of actomyosin has been realized since the work of Weber and Weber. They used glycerinated fibers and the phenomenon was further studied by Hasselbach and Weber. Recently, Perry and Grey have turned attention to this effect of overoptimal ATP concentration in connection with the mechanism of relaxation.

Using washed fresh myofibrils and working in the range of 2.5 to 10 × $10^{-3}M$ for both ATP and Mg^{++}, they found that the ATP became inhibitory when its molar concentration exceeded that of the Mg^{++}. This relationship depended to some extent on the previous history of the myofibrils; it is well seen with myofibrils freshly isolated in 0.1 *M* KCl, and tested in a medium of about physiological ionic strength, 0.16. It does not, however, hold at lower concentrations of the reactants. Thus Geske*et al.* have found that with Mg^{++} concentrations below 5 × $10^{-4}M$, the overoptimal ATP concentration shifts (with increasing Mg^{++}) first to *lower* ATP concentrations.

Perry and Grey found that marked substrate inhibition is characteristic of Mg^{++}-activated ATPase; it is very slight with Ca^{++} as activator. Very low Ca^{++} concentrations can abolish the substrate inhibition of the Mg^{++} activated system.

CONTRACTION AND ATPASE ACTIVITY

Weber and his collaborators have brought forward much evidence showing the close correlation between tension production in glycerinated fibers and their ability to dephosphorylate ATP. Thus Weber and Weber showed the effect of changing ATP concentration upon both tension and ATPase activity; the changes ran parallel both at room temperature and at 0°C.

Ulbrecht and Ulbrecht, with glycerinated fibers from smooth adductor muscles of *Anodonta,* found that the mechanical effects of ATP application and the ATPase activity showed the same temperature dependence. Bendall, working on the relaxing factor, found that as his preparations decreased the rate of shortening with glycerinated fibre fundles, so also they caused lowered ATPase activity. In all these cases, Mg^{++} was used as the ATPase activator and was supplied in the medium of the contracting fibers. Bowen and his collaborators, on the other hand, have described a number of cases which they consider can better be explained by correlating combination of ATP and actomyosin (rather than ATPase activity) with contraction. Thus they found conditions in which addition of Mg^{++} ions accelerated shortening of actomyosin filaments in presence of ATP, while homogenized preparations from the same material showed unchanged or decreased ATPase activity. On the other hand, Ca^{++} ions, which increased the enzymatic activity by a factor of 4, decreased the rate of shortening.

At this point, it is important to remember that the phenomenon of contraction is complex: ATP splitting may be an essential feature, but those who take this point of view would not regard it as the only essential requirement. Thus there seems good reason to believe that Mg^{++} ions play an important role (which cannot be filled by Ca^{++} ions) both in viscosity and in gelation changes; that they should play a specific part in the shortening process is therefore not unexpected.

Indeed, Bendall (personal communication) has found, with well-washed glycerinated fibre bundles (about 200 μ in diameter) shortening under load, that there was virtually no work performance in 6 m*M* ATP when Ca^{++} was added up to 8 m*M*, although this Ca^{++} concentration can stimulate about maximal ATPase activity. Addition of 0.4 m*M* Mg^{++} to the bath (with or without Ca^{++}) led to good work production and ATPase activity. It seems, then, that the correlation is between Mg^{++}-activated ATPase activity and contraction.

Again, increasing the KCl concentration in the presence of Mg^{++} ions leads to accelerated shortening, while the enzymatic activity is depressed. Here, as Perry has pointed out, the increased ionic strength may, by reducing hydration of the fibers, improve their mechanical properties, quite independently of any effect on enzyme activity.

Further, it has been found that the final degree of shortening of glycerinated fibre bundles varied with the concentration of ATP supplied, reaching in each case a constant value. The conclusion was drawn that the degree of shortening did not depend on the amount of ATP split, but on the concentration present, i.e. on the amount bound to the protein. But in these circumstances, the limiting factor in the hydrolysis rate in the interior of the fibers would be the rate of diffusion, itself dependent on the concentration.

There is nothing in these data to show that the shortening did not depend on the splitting rate; this is admitted but it is maintained that the comparison of the effects of Mg^{++} ions with those of Ca^{++} ions rules out this possibility. However, we have already shown the reasons for regarding this last argument as unsatisfactory.

Many reagents reacting with SH groups are known to inhibit the ATPase activity of myosin. Their effect on contraction of actomyosin threads or fibers has always been found to be similarly or even more strongly inhibitory. In some cases, the ATPase activity of fibers after partial poisoning may be higher at room temperature than that of the unpoisoned fibre at 0°C.; yet the latter will contract, the former not. It would seem that, in the former, the energy released is not available for contraction.

It may be that SH groups are essential for reaction along the protein chain as well as for ATPase activity. To sum up, we may say that the evidence strongly favors the necessity of participation of Mg^{++}-activated ATPase in contraction; but the contraction process is complicated, involving more than this enzymatic activity.

MECHANISM OF THE ACTOMYOSIN ADENOSINE TRIPHOSPHATE INTERACTION IN CONTRACTION

We must now consider the stages of this interaction in more detail in relation to the mechanism of contraction. Ever since the realization of the part played by ATP, opinion has been divided as to whether its phosphate was liberated during the contraction or the relaxation phase; we shall deal with arguments put forward on both sides.

It must be remembered that, until a few years ago, the view was generally held that contraction was produced by the folding of protein chains which ran throughout the muscle length. Recently, much evidence has accumulated that striated muscle contains two types of filament and that the individual filaments do not change in length as the muscle shortens. From these observations, the idea follows that the two types of filament slide over one another to produce the shortening. It has been suggested that a cyclic mechanism underlies this relative movement, linkages between the two types of filament (actin and myosin) being made and broken during contraction as the filaments pass each other.

THEORIES INVOLVING THE LIBERATION OF FREE ENERGY

Szent-Györgyi has for many years been a strong protagonist of this point of view, though it would seem that he has recently joined those who regard the contraction phase as the one needing direct provision of energy by ATP breakdown. He had supposed that upon stimulation, actin plus myosin forms actomyosin, and that ATP then combines with the myosin. In doing so, it enables a new endergonic link to be formed at the expense of the energy-rich phosphate bond. The establishment of this link has made certain joints pliable and these now fold, as a result of electrostatic attractions and repulsions along the chain. In order that relaxation may take place, the new energy-rich link must be broken, and this is done by the removal of ATP from the protein as ADP + H_3PO_4.

Avariation of this view is that of Riseman and Kirkwood; they postulated the phosphorylation of OH groups along the protein chain, which was then held extended by repulsion of the negative charges. On stimulation, dephosphorylation-took place, the charges were thus abolished, and the chain was free to contract. In relaxation, rephosphorylation of OH groups by ATP was necessary. On this theory, inorganic phosphate would actually be released during contraction, but the loss of free energy by the breakdown of ATP and formation of the phosphate ester linkage would occur during relaxation.

Next we come to the views of Morales and his

collaborators, put forward during the last few years and summarized by Morales*et al.* These studies have been made on lightscattering by solutions of myosin B, i.e. the protein obtained by 5 to 24 hours' extraction, and considered by most workers to consist of a mixture of actomyosin and myosin, the former in much greater amount.

Morales and his co-workers find the evidence for the presence of actin inconclusive, and consider the lightscattering phenomena to be due to myosin alone. The lightscattering method, as we have seen, is a very sensitive and rapid one for following changes in shape and size of protein particles. The effect of added ATP is a fall in the light scattering which persists for a time (during dephosphorylation of the ATP) and then there is a return to the original value. Below a certain ATP concentration, the degree of change depends on the concentration of the ATP added.

Further when the experiment was carried out in presence of 0.001 *M* Mg^{++} (which inhibits dephosphorylation at this high ionic strength), the lightscattering change (like the corresponding viscosity change) was unaffected in degree, but now was not reversed. This was interpreted to mean that the deformation of the particles depended on their combination with ATP and went on irrespective of ATP breakdown.

$$E + ATP \rightarrow E\ ATP \rightarrow E + ADP + P$$

where *E* represents the enzyme after deformation.

By studying the reaction at different temperatures, Morales and his collaborators consider that they have evidence that the light-scattering changes can go on without a change in molecular weight of the particles—i.e. without dissociation. In this, although using the same Zimm method, they are at variance with Gergely. In later work evidence for some dissociation was found by the Morales group, but the authors consider that only different states of aggregation of myosin itself were concerned when this happened.

Gergely and Martonosi on the other hand, have obtained further evidence in support of their interpretation, that dissociation of actomyosin is concerned, by preparing typical actin from 3 to 6 times precipitated myosin B. Morales and his collaborators concluded that the combination between protein and ATP is an energy-providing reaction (ÄF° = —6600 cal. per mole). They consider this deformation of the protein particles to be responsible for contraction and to depend upon the charge. They picture the degree of deformation as governed by a conflict between extensile electrostatic forces and contractile entropic forces—the particles gain configurational entropy on shortening, whether the shortening is by random coiling or by folding into an ordered form by reaction between certain groups. For relaxation, an energy-providing reaction is needed, to overcome the electrostatic effects of the bound ATP, e.g. the removal of the ATP from the protein by enzymatic breakdown to ADP and inorganic phosphate. The experiments that we have already described in which ATPase activity and contraction could be to some extent affected independently are cited as evidence for this point of view.

There is indeed abundant evidence that combination between ATP and myofibrillar protein in solution takes place and that "deformation" of the protein occurs before ATP dephosphorylation. Other workers have obtained results by the lightscattering method similar in many respects to those just described. The experiments employing the viscosimetric method cited earlier in Section II, B indicate the same thing. However, there seems to be, as we found in considering the effect of ATP on actomyosin solutions, good evidence that the "deformation" is a dissociation of myosin from actin.

When the effects upon a sol were compared with those on various forms of gel, we regarded the initial increase in extensibility as a comparable dissociation, a preliminary loosening of linkages putting the rigid gel into a fit state for contraction. In the resting muscle, as we shall shortly see, this step is unlikely to be necessary, since electron micrographs seem to show the actin and myosin as distinct and separate filaments. But recent work has made it likely that any contraction involves a cyclic process of linkage-making and linkage-breaking, so that this dissocia-tion stage would be involved during the contraction phase (to make possible each further shortening step) and during relaxation.

FREE ENERGY OF ATP HYDROLYSIS

We have already discussed a good deal of experimental work which, taken at its face value, would seem to support this point of view. However, it is obvious that no enzymatic action by the actomyosin upon the ATP can take place without previous combination of the two, so that there is a logical flaw in any argument from a correlation between tension production or shortening and ATPase activity.

The most striking evidence in support of this standpoint—that the energy of ATP hydrolysis is needed in the contraction phase—is that brought forward by H. H. Weber and by Bozler. We have already seen that a glycerinated fibre, made to contract in ATP solution, and remaining rigid and contracted when excess ATP is washed away, can be made to show a momentary relaxation by the addition of more ATP.

Weber and his collaborators now showed that this relaxation can be brought about in circumstances al- lowing no simultaneous energy provision. Thus ATP can bring about complete and lasting relaxation in the presence of the mercurial Salyrgan, which prevents its hydrolysis by poisoning the actomyosin ATPase. Salyrgan alone has no effect on the fibre. Inorganic pyrophosphate (0.0 15 M) also gave the same effect, although it is not a substrate for actomyosin ATPase. Bozler also got relaxation of ATP-contracted glycerinated fibers when 0.02 *M* pyrophosphate was added, and concluded that the relaxation process involves no large energy change. Bendall has got similar results with much lower pyrophosphate concentration (0.004 M), provided Mg^{++} (0.004 *M*) was also added.

This view of the timing of energy provision accords with A. V. Hill's finding that the relaxation phase (provided the muscle relaxes unloaded) is free from heat production.

H. H. Weber has put forward a scheme which enables one to visualize possible mechanisms whereby the free energy of hydrolysis could be directly used in the contraction of the protein molecule.

Schemes such as this must inevitably be open to criticism as regards their details. Thus relaxation is pictured as brought about by hydrolysis of the ester linkage, and it is difficult to see how the relaxing effect of ATP fits in here. Morales and Botts have pointed out that the forces leading to establishment of covalent linkages are of very short range; further, such forces would become larger as the muscle shortened, while in fact the isometric tension developed becomes less as the muscle shortens. These latter objections are to some extent overcome when some such mechanism is envisaged (Weber, 1958) acting as part of a cyclic process. The evidence concerning the possibility of phosphorylation has already been considered.

THE SLIDING HYPOTHESIS

Electron micrographs of striated muscle have shown that two sets of filaments are present in regular hexagonal array and parallel to the long axis. Of these, the larger (diameter about 100 A.) have been identified as consisting mainly of myosin, the more slender (40 A. in diameter) as mainly of actin. The myosin filaments occupy the A band; the actin filaments run from the Z lines at the centre of the I band and into the A band as far as the beginning of the H zone. Upon contraction (to 6570% of the resting length), the A band remains of constant length, but the H zone and the I band shorten. These changes are best explained by a sliding of the actin filaments past the myosin into the A band, without folding of either type of filament unless shortening of the muscle is very great.

The evidence from low-angle diffraction diagrams obtained with living muscle supports the conclusion that the filaments themselves do not change in length. The problem then is to discover the mechanism whereby the actin filaments are drawn past the myosin filaments.

In most of the suggested hypotheses, the contraction proceeds by stages. Sites on the actin filament can be pictured as oscillating by means of thermal agitation backwards and forwards past sites on the myosin filament with which it overlaps. As a result of the stimulus, linkages are formed between the two sites and a certain degree of contraction occurs. When these links are broken by the arrival of fresh ATP, the actin sites are now within range of further myosin sites and the process is repeated further and further along the myosin chain. A schematic formulation is given below, based on what we know from studies *in vitro* of the interactions of actin, myosin and ATP.

1	(During the period of active contraction)	M.ATP $\rightarrow$ P + ADP	
2	(Formation of links, shortening stages)	P + A $\rightarrow$ AM + P +	free energy used as tension or work
3	(Loosening of links, finally relaxation)	AM + ATP $\rightarrow$ M. ATP + A	

Reaction: It is supposed that the effect of the stimulus is to cause a transphorylation which provides one of the proteins (myosin is suggested here but it might be actin) in a state ready to combine.

Reaction: The free energy of the energy-rich bond is used in some reaction or series of reactions leading to the formation of contracted actomyosin. This series of reactions might be, for instance, along the oblique connecting side chains suggested in one hypothesis or occupy the place of the "spring" in the hypothesis of A. F. Huxley. Formation of covalent bonds might be concerned as in Weber's formulation; electrostatic forces or hydrogen bonds might be concerned—we do not know.

Reaction: The loosening of the linkages would be due to the arrival at the sites of fresh ATP, by diffusion.

On this formulation, ATP is concerned during contraction with both making and breaking of links; during relaxation, only with breaking of links. There is no requirement for energy provision during relaxation. The activation heat of Hill could arise:

- Possibly in connection with unknown reactions between stimulation and reaction (1);
- Possibly from some ATP dephosphorylation in the neighborhood of the sites, necessary to reduce it below a critical concentration;
- Perhaps from some heat wastage in reaction (1);

- From heat production connected with reaction (2), before enough links are formed to permit a mechanical response.

The shortening heat, only produced when the muscle actually shortens, would be derived from reaction (2), possibly also from reaction (3), though the latter contribution would have to be small as no heat is detectable during relaxation without load.

A very striking aspect of muscle metabolism is the constancy of the heat of shortening for a given muscle. Whether the muscle contracts slowly under a load or rapidly unloaded, the extra heat associated with a given amount of shortening is the same. The energy necessary for the accomplishment of the work is specially mobilized in proportion to the load and the shortening heat is not diminished to provide any part of it. Now as the muscle contracts more slowly when it is loaded, in these conditions more time is available and the same links might be formed and broken several extra times. We may surmise that this is the source of the extra energy which appears as work. Bendall has recently drawn attention to an assumption which is necessary if the sliding hypothesis is to be reconciled, as it must be, with the proportionality between shortening heat and distance shortened. It seems that as the actin and myosin filaments slide over each other, the number of reacting groups per unit of shortening must remain the same. Two alternative possibilities present themselves:

- On the actin filaments only the tips are reactive and groups situated here react with successive groups on the myosin;
- On the myosin filaments only the ends near the I band bear active groups and these react with site after site on the actin filaments. In either case a biochemical problem is presented, since so far as is known each type of filament is homogeneous along its length. Bendall makes the tentative suggestion that tropomyosin, quantitatively a minor constituent of the myofibril may have the rôle of protecting some parts of the major proteins from interaction.

The effort to go deeply into the sliding mechanism is only just beginning. A. F. Huxley has suggested a very interesting picture, involving oscillation of the myosin sites about an equilibrium position in the neighborhood of the actin sites, with certain postulations about the rate constants of the making and breaking of links, and the factors affecting these constants. Very reasonable quantitative agreement was obtained between the expectations worked out mathematically from this hypothesis and a great number of the thermal and mechanical observations on living muscle recorded in the literature.

ENERGY PROVISION IN THE LIVING MUSCLE

Using evidence from innumerable observations and experiments *in vitro,* we have built up the picture of an actomyosin-ATP machine depending for its energy supply on ATP dephosphorylation. We wish now to inquire how well this picture fits the requirements set by phenomena *in vivo* as regards the arrangements for the fuel supply.

That is to say, how good is the evidence that ATP breakdown is *in vivo* the primary energyyielding reaction? It is not to be expected that, with a fresh, unfatigued muscle, such evidence will easily be produced, since, as we have seen, there are powerful enzyme systems in the muscle forrephosphorylation of ADP. Nevertheless, a number of lines of inquiry can be explored.

ATPASE ACTIVITY

In the first place, we may ask whether the activity of actomyosin ATPase under conditions *in vivo* is high enough to account for the energy production if all this is channeled through ATP breakdown. It seems that we must consider the behaviour of actomyosin ATPase, not that of free myosin ATPase. Muscle contains both Mg^{++} and Ca^{++}, but there is good evidence that only Mg^{++} activated actomyosin ATPase is concerned with contraction.

The results of Perry and Grey give a figure of about 0.3 μ, mole per milligram protein per minute for the Mg^{++}-activated

ATPase of well-washed rabbit myofibrils at 20°C. The figures of Hasselbach are similar for finely-divided actomyosin gel at 23°C. Thus a figure of I to 2 μ moles per milligram protein per minute might be expected at 37°C., or 1 to 2 × 120 μ moles per gram muscle. This assessment takes no account of the high initial rate of inorganic phosphate liberation, which may be much more than double the rate estimated by observation over several minutes. If this high rate is substantiated as due to true ATPase activity, we must suppose it would operate in the conditions of discontinuous ATPase activity we have been contemplating. With 120 mg. of actomyosin per gram of muscle, the rate would then be at least 2 to 4 × 120 μ moles, or 2.4 to 4.8 × 10-4 moles per gram muscle per minute.

Bendall (personal communication) has indeed found initial values at 35°C. for rabbit myofibrils corresponding to 5-6 × 10-4 moles per gram muscle per minute. Mommaerts had calculated the utilization of energy-rich phosphate which would correspond to the maximum effort in human muscle; these calculations are based on the extra oxygen consumption, on the assumption that 5-6 energy-rich phosphate bonds are formed per molecule of oxygen used, and include corrections to allow for the actual amount of muscle tissue involved in the increased metabolism.

It seems that there may be a 500-fold rise in metabolism, and the metabolism of the energy-rich phosphate bond should reach 10 × 10-4 moles per minute per gram of muscle. The comparison of ATPase activity of rabbit muscle preparations with the muscle metabolism of man is open to criticism, since the small mammals show greater metabolic activity than the large. However, considering the roughness of all these calculations, the measure of agreement between the need and the supply is reassuring.

pH CHANGES IN LIVING MUSCLE

In experiments of Dubuisson, a glass electrode (timelag 2 to 3 sec.), in close contact with the muscle, was used to examine pH changes consequent upon a 4 sec. tetanus in the frog gastrocnemius.

A series of three pH changes was seen-towards the acid side during shortening; then towards the alkaline side; finally another change to the acid side supervened. The last two changes were mostly postcontraction. Dephosphorylation of ATP sets free acid, while dephosphorylation of creatine phosphate sets free base. The conclusion that the first change is due to ATP breakdown, the second to creatine phosphate breakdown, and the third to lactic acid formation, was supported by the effects of varying the initial pH of the muscle and of poisoning with iodoacetate.

Chapter 11

DNA as Genetic Molecule

Chromosomes were known to be made up largely of protein and nucleic acid, the latter of the deoxyribose type. Deoxyribonucleic acid has now become widely known as DNA. Are the genes protein of DNA—or possibly both? At first their specifications were thought to reside in protein, for chemists knew that proteins are long polymer molecules consisting of linear sequences of amino-acid building blocks of some twenty kinds.

Since there are obviously almost unlimited possibilities in proportions and sequences of amino acids, it was easy to believe that the gene was a kind of coded message in protein. DNA, on the other hand, was believed by many to be a rather monotonous polymer built of four kinds of nucleotide units arranged in segments of four that were repeated many fold. If this were its structure, there would clearly be little opportunity for specificity; therefore it was not considered a serious candidate for the role of primary genetic material.

A series of investigations on pneumococcal bacteria, beginning in 1928 and leading up to the classical paper of Avery, MacLeod and McCarty, severely shook the faith in protein as the basic stuff of heredity. It became clear that the type of polysaccharide capsule produced by a pneumococcal strain could be altered experimentally by treating it with pure DNA from another strain.

Thus DNA from a type III pneumococcus could cause a type II recipient cell to be transformed permanently into type III, and henceforth to produce DNA specifying type III polysaccharide.

The experimental procedure by which this is accomplished is not quite as simple as the above account implies. Nevertheless it appeared that polysaccharide-specifying DNA from the donor strain might be entering the recipient and somehow replacing its homologous DNA. There were, however, alternative interpretations, which, though less straightforward, continued to be preferred by many geneticists and biochemists.

By 1953 more direct evidence had come from another source. By then the life-cycles of certain bacterial viruses—bacteriophage—had been worked out and sufficient genetic work done on them to make it clear that like higher organisms they exhibited particulate inheritance. In other words, these viruses, too, have genes.

Bacterial viruses consist largely of protein coats containing cores of DNA. Electron microscopy suggested that when a bacterial cell is infected by a virus particle, its coat does not enter the host cell. The matter was most elegantly settled by the use of radioactive tracers. Hershey and Chase infected bacteria with viruses whose protein coats were labelled with sulphur-35, a radioactive isotope. The labelling was done by growing a crop of viruses on bacteria that in turn had been grown on a medium containing radioactive sulphur in the form of sulphate.

After infection, the protein coats were removed from the bacterial hosts by shearing them off in a high speed blender. Separated by centrifugation, the virus coats and infected host cells could be separately examined for presence of radioactive sulphur. Almost all of it was found in the virus coats that did not enter host cells. Since proteins contain sulphur while DNA does not, this result suggests that only DNA enters the host on infection. Experiments in which DNA was labelled with radioactive phosphorus-32 led to the same conclusion.

Now the coats were largely unlabelled after infection had occurred, while the host cells contained most of the radioactivity. Since DNA contains phosphorus—one atom per nucleotide—and protein does not, it is clear that DNA and not much else enters the host cell. Since viruses have genes and only DNA enters the host cell, the viral genes must be DNA. Actually a small amount of protein does enter with the DNA, but labelling experiments in which the radioactivity entering a host cell in protein or DNA is followed to the next viral generation show that it is the DNA, not the small amount of protein, that is responsible for the transfer of genetic information from one generation of viruses to the next.

What about higher organisms—spinach plants and man? Are there primary genetic specifications likewise written in DNA? The evidence is not conclusive but since the most conservative hypothesis assumes they are, we proceed on that assumption until evidence to the contrary comes forth.

STRUCTURE OF DNA

What is the structure of DNA that enables it to carry hereditary information? A tremendous step was taken toward answering this question in 1953 by a young American biologist, James D. Watson, and the English chemist, Francis H. C. Crick, working together at Cambridge University.

According to this model, DNA consists of a pair of antiparallel polynucleotide chains wound helically around a common axis and cross-linked through specific hydrogen bonding between purine and pyrimidine bases. Letting A, T, C, and G represent the four nucleotides characterised by adenine, thymine, cytosine and guanine, a four-unit segment of DNA can be represented in two dimensions as follows:

A-T-C-G
....
T-A-G-C

Paired dots represent hydrogen bonds. The two chains of the DNA molecule run in opposite directions. This is determined by the orientations of nucleotides in the two chains.

INFORMATION IN DNA

Genetic information must somehow be determined by sequence of nucleotides. Since only A:T and C:G nucleotide pairs are possible in the normal structure, the two chains obviously carry complementary information. As we shall see this is of special significance in hypotheses of DNA replication.

One can think of information being carried either as a sequence of nucleotide pairs in the double molecule or as nucleotide sequences in the two single chain components.

How much DNA is in a single human cell? It is estimated that the total DNA of the 46 chromosomes of a fertilised human egg contains something like 5,000,000,000 nucleotide pairs. Since the genetic material is carried in duplicate in such a cell, one complete set of information is written in sequence of some 2,500,000,000 such pairs.

Prof. Crick has estimated that this amount of DNA is sufficient to, Making use of the information then available about DNA—nucleotide composition of DNA of various sources, X-ray diffraction observations of M. H. F. Wilkins and associates at King's College, London, general knowledge of the structural arrangements of atoms in nucleotide components, etc.—they succeeded in constructing a molecular model of DNA that seemed to satisfy essentially all requirements. Now, some seven years later, it is generally agreed that the Watson-Crick structure is essentially correct for the native DNAs of a number of organisms. It is now clear that their achievement is outstanding in modernday biology. This is so because their model goes so far in suggesting plausible answers to the questions posed at the beginning of this lecture.

According to this model, DNA consists of a pair of antiparallel polynucleotide chains wound helically around a common axis and cross-linked through specific hydrogen bonding between purine and pyrimidine bases. Letting A, T, C, and G represent the four nucleotides characterised by adenine, thymine, cytosine and guanine, a four-unit segment of DNA can be represented in two dimensions as follows:

A-T-C-G

....

T-A-G-C

Paired dots represent hydrogen bonds. The two chains of the DNA molecule run in opposite directions. This is determined by the orientations of nucleotides in the two chains.

INFORMATION IN DNA

Genetic information must somehow be determined by sequence of nucleotides. Since only A:T and C:G nucleotide pairs are possible in the normal structure, the two chains obviously carry complementary information. As we shall see this is of special significance in hypotheses of DNA replication.

One can think of information being carried either as a sequence of nucleotide pairs in the double molecule or as nucleotide sequences in the two single chain components.

How much DNA is in a single human cell? It is estimated that the total DNA of the 46 chromosomes of a fertilised human egg contains something like 5,000,000,000 nucleotide pairs. Since the genetic material is carried in duplicate in such a cell, one complete set of information is written in sequence of some 2,500,000,000 such pairs.

Prof. Crick has estimated that this amount of DNA is sufficient to encode the contents of some 500 large library volumes. This is another way of saying that the genetic specifications for producing a person from an egg cell, given a proper environment, adequate food of the right kind, etc., might be written in English in this number of volumes.

What about the physical size of DNA molecules? The diameter of the double helix is 20 Ångström units. The base pairs are spaced at 3·4 Å intervals. Thus a continuous linear double helix of the 5,000,000,000 base pairs of a single human egg would be somewhat more than 5 ft. in length —only slightly less than the height of an average person. This is a convenient way to remember the amount of DNA in a single human cell. If the strands were packed side by side in a monolayer on the head of a pin 1 mm. in diameter, this amount of DNA would cover less than 1/200 of the surface.

REPLICATION OF DNA

The Watson-Crick structure immediately suggested how DNA molecules might reproduce by separation of the double structure into single chains followed by sythesis of new partners against each of the old chains. Since the two originals are complementary, each carries the information necessary to direct the synthesis of its partner. It is presumed that single chains serve as templates against which free nucleotides are properly ordered by specific hydrogen pairing. Just how the problems of separation of paired chains helically coiled around a common axis is solved is not known. The forces required to "untwist" the double helix would not be great if it were to rotate around its axis in the manner of an automobile speedometer cable.

Is there any evidence that DNA replication in fact occurs in this way? Yes, two kinds of evidence indicate that it does.

One of these depends on labelling old and new chains so that they can be distinguished. This can and has been done with radioactive phosphorus-32, but there are difficulties in this method that have not been overcome in an entirely satisfactory manner.

A second method of labelling involves the use of the stable heavy isotope of nitrogen, N15. Bacteria grown on a culture medium containing only N15 eventually become fully labelled with this isotope. Since there are seven or eight nitrogen atoms per pair of nucleotides of molecular weight approximately 700, replacing all N14 with N15 increases the weight a bit over one per cent. Since the size does not change, the density increases by a corresponding amount. "Heavy" DNA molecules can be separated from the light variety in an analytical ultracentrifuge.

The method for doing this was first developed by Meselson Stahl and Vinograd and consists in centrifuging DNA in a cesium chloride solution of proper density. The cesium chloride molecules are thrown down in a high centrifugal field, but, being small, they diffuse sufficiently to establish an equilibrium density gradient in the centrifuge cell. If the range of density so established includes that of DNA, the DNA molecules in solution will form a band at a level exactly corresponding to their bouyant density. Being large they diffuse only slowly and hence form a narrow band, the position of which is easily established by means of an ultraviolet optical system. A mixture of N15 and N14 DNA molecules form two cleanly separated bands.

This method makes possible the following elegant experiment first carried out by Meselson and Stahl. Bacteria are grown on N15 medium until equilibrated, that is, until they become uniformly heavy. The bacteria are then transferred to a medium containing only nitrogen of atomic weight 14. After one cell generation, as determined by doubling of the total population, all DNA molecules should have one heavy old chain and one light new complement.

That is, they should be "hybrid" and hence intermediate in density between heavy and light DNA. They are. In a second round of replication—which doubles the population again-heavy and light chains should separate and each then directs synthesis of a light complement. Hence, half the DNA molecules after exactly two cell divisions should be hybrid and half light. Again they are.

If a population of hybrid DNA molecules—present after one replication of heavy molecules in an N14 medium—are heated under the right conditions, molecules about half the molecular weight are found and they are light and heavy in equal numbers. The evidence is fairly convincing that these are single chains of DNA.

This experiment does not prove that the Watson-Crick hypothesis of DNA replication is correct but it strongly suggests it. A perverse nature might have devised another way of giving the observed result.

THE KORNBERG SYSTEM

An even more dramatic way of investigating the mechanism of DNA replication is that devised by Arthur Kornberg and his co-workers. In a suitable buffer solution containing magnesium ions, the four nucleotides of DNA as triphosphates, and a DNA polymerising enzyme, DNA is rapidly synthesised if primer DNA molecules are added. Single stranded DNA, obtained by heating native DNA, is much more effective as a primer than is carefully prepared native material. Something like a twentyfold increase in DNA over that added as primer has been obtained.

That the primer is copied is indicated by the fact that the base and Vinograd (1957) and consists in centrifuging DNA in a cesium chloride solution of proper density. The cesium chloride molecules are thrown down in a high centrifugal field, but, being small, they diffuse sufficiently to establish an equilibrium density gradient in the centrifuge cell. If the range of density so established includes that of DNA, the DNA molecules in solution will form a band at a level exactly corresponding to their bouyant density.

Being large they diffuse only slowly and hence form a narrow band, the position of which is easily established by means of an ultraviolet optical system. A mixture of N15 and N14 DNA molecules form two cleanly separated bands.

This method makes possible the following elegant experiment first carried out by Meselson and Stahl (1958). Bacteria are grown on N15 medium until equilibrated, that is, until they become uniformly heavy. The bacteria are then transferred to a medium containing only nitrogen of atomic weight 14.

After one cell generation, as determined by doubling of the total population, all DNA molecules should have one heavy old chain and one light new complement. That is, they should be "hybrid" and hence intermediate in density between heavy and light DNA.

They are. In a second round of replication—which doubles the population again-heavy and light chains should separate and each then directs synthesis of a light complement. Hence, half the DNA molecules after exactly two cell divisions should be hybrid and half light. Again they are.

If a population of hybrid DNA molecules—present after one replication of heavy molecules in an N14 medium—are heated under the right conditions, molecules about half the molecular weight are found and they are light and heavy in equal numbers. The evidence is fairly convincing that these are single chains of DNA. This experiment does not prove that the Watson-Crick hypothesis of DNA replication is correct but it strongly suggests it. A perverse nature might have devised another way of giving the observed result.

THE KORNBERG SYSTEM

An even more dramatic way of investigating the mechanism of DNA replication is that devised by Arthur Kornberg and his co-workers (1959). In a suitable buffer solution containing magnesium ions, the four nucleotides of DNA as triphosphates, and a DNA polymerising enzyme, DNA is rapidly synthesised if primer DNA molecules are added. Single stranded DNA, obtained by heating native DNA, is much more effective as a primer than is carefully prepared native material. Something like a twentyfold increase in DNA over that added as primer has been obtained.

That the primer is copied is indicated by the fact that the base composition of the product is like that of primer. DNAs from different sources may have quite different rations of A:T to C:G base pairs, and it is therefore possible to determine whether the ratio of the primer is reproduced in the product. Without primer, DNA is spontaneously synthesised in the Kornberg system after a lag of two to four hours. But unlike natural DNA, this spontaneously synthesised material contains only A:T base pairs. The As and Ts alternating in sequence in each of the two chains as follows:

If this A:T copolymer is now used as a primer in a fresh system, more A:T polymer is formed without a lag period. C and G nucleotides are excluded although present in abundance in the system. Here too it appears that the primer is copied as postulated. Again, the agreement between hypothesis and the facts observed in the Kornberg in vivo synthesis of DNA does not prove the hypothesis. But it enormously strengthens the case.

USE OF DNA INFORMATION

How is information in the form of DNA made use of in the development of an organism like man? This is clearly a question of the most fundamental importance to biology. At the same time it is an enormously difficult one and we are a long way from knowing the complete answer.

All living systems contain DNA (or in some viruses a related form of nucleic acid, ribonucleic acid, called RNA) and protein. In cellular forms the proteins are of many kinds. We know that many genes, perhaps all, are somehow concerned with synthesis of specific proteins.

Many of these serve as organic catalysts—enzymes—or as components of these catalysts. Enzymes accelerate vital reactions that would otherwise proceed at rates too low to sustain life. In cellular organisms there are thousands of kinds of enzymes, each owing its specificity to a particular protein. We can therefore narrow the problem of gene action, at least in some cases, to that of protein synthesis.

An hypothesis at present widely used as a working basis visualises the process of protein synthesis in the following way: for each protein potentially capable of being formed, there is in the nucleus a specific segment of DNA that carries the information by which the twenty kinds of amino acid sub-units in that protein are properly ordered during its synthesis. Take human haemoglobin as an example. Each molecule of this vital oxygen-carrying red protein consists of four protein chains and an equal number of heme groups each containing an atom of iron.

The protein chains are of two kinds, called alpha and beta chains. There are two of each, the members of a pair being identical and each made up of about 150 amino-acid units arranged in a precisely determined sequence. For each of the two kinds of protein chains, there is presumed—with some evidence—to be a gene consisting of a segment of DNA. This segment is, by the definition I shall use, a gene. It may be something of the order of 1000 nucleotide pairs in length.

How is the information in the gene for the alpha haemoglobin chain made use of? The hypothesis assumes the following sequence of events: from the gene in the nucleus, information is transferred to RNA, presumably by the DNA somehow acting as a template in the ordering of the four kinds of ribose nucleotides in RNA.

This informational RNA then moves from the nucleus to the cytoplasm of the young red blood cell—before it loses its nucleus. There it is incorporated into microsomes, which are sub-microscopic bodies made of structural protein and structural RNA. Once in microsomes, informational RNA molecules serve as templates against which amino-acids are arranged in proper order to make alpha haemoglobin protein chains.

Prior to this the amino-acids are activated and attached to small carrier segments of RNA, each specific for its own amino-acid. Thus for each of the twenty amino-acids there is a corresponding carrier RNA segment. Carrier RNA molecules are somehow coded to specific sites on the informational RNA in the microsome. In this way each amino-acid is carried to its proper position on the template. They are then joined through peptide linkages to form alpha chains and are released from the template and microsome. Whether association of alpha and beta chains with their hemes, and with each other, occurs in the microsome or outside is not known.

Not all proteins for which genetic directions are available in the nucleus are synthesised in any one cell. There are ingenious control mechanisms in operation which determine whether the information in a given gene will be used and, if so, when and for how long. In some cases it is clear that the presence or absence of substrate determines whether a given enzyme will be synthesised. In this way it is assumed that enzymes are not made in quantity when there are no substrates on which they can work. Through studies on such control mechanisms we are beginning to understand how it is that cells with identical genetic information in their nuclei may do quite different things, depending on their environmental context or on their previous history.

There are now known many instances in a variety of organisms in which specific protein variation is known to be related to particular genes. I have already referred to haemoglobin protein in man. Similar situations are known in viruses, bacteria, algae, fungi, insects, higher plants, mammals, etc. The haemoglobin case in man is interesting in that it is known that gene changes may result in substitution of a single amino-acid in a chain of 150 units.

Thus sickle cell haemoglobin, synthesised under the direction of a modified form of the gene in control of the beta protein chain, differs from normal haemoglobin in that a single glutamic acid unit in the chain is replaced by a valine. In the presence of a third form of this same gene the same glutamic acid unit is replaced by the amino-acid lysine. In all, about a dozen modifications of human haemoglobin are known, many of them investigated genetically.

Another example in man is found in the genetic disease galactosaemia. Here is a specific enzyme, galactose-1-phosphate uridyl transferase, essential for the conversion of galactose into a usable form, which is partially or wholly inactive. The enzyme defect is referable to a gene change.

In bacteria, fungi and other micro-organisms correlations between gene change and specific protein modification are much easier to detect than they are in man with his several obvious disadvantages for biochemical and genetic investigations. As a result there are dozens of examples known in such forms as Escherichia, Salmonella and Neurospora.

MUTATION

In terms of DNA structure gene mutation is believed to be the result of alterations in nucleotide sequence. Many mutations are thought to be the result of errors in DNA replication. In their original paper on DNA structure, Watson and Crick pointed out that if at the precise moment of partner selection a purine or pyrimidine were to exist in an improbable tautomeric form, it might form hydrogen bonds with a "wrong" nucleotide.

At the next round of replication such a "wrong" nucleotide would direct that its own complementary nucleotide be inserted in the paired chain. Thus one of the two daughter DNA molecules would differ from its sister in one nucleotide pair. Its informational content would be modified accordingly.

Experimentally mutations can be increased in frequency in a number of ways. High-energy radiations—ultraviolet, X-rays, etc.—are effective in doing this. Transmutation of artificially incorporated radioactive isotopes such as phosphorus-32 is known to produce mutations. Since radioactive isotopes occur naturally in low frequencies, they are no doubt responsible for a small fraction of spontaneously produced mutations.

Many chemical substances are known to be mutagenic. Perhaps the one best known from the standpoint of mechanism of action is nitrous acid. This specifically oxidises amino groups. In this way it may chemically change a natural DNA pyrimidine or purine in such a way that its hydrogen bonding specificity is modified. The mutagenic properties have been especially well studied in a tobacco mosaic virus, an RNA virus. Here the natural pyrimidine, cytosine is converted to uracil, which is also a natural occurring pyrimidine in RNA. Adenine is oxidised to hypoxanthine, the latter not a normal constituent of RNA. In a similar way guanine is changed to xanthine. It appears that the oxidation of any one of 3000 of the 6000 nucleotides in a tobacco mosaic virus particle results in an inactivating mutation.

The purine and pyrimidine analogues, 2-aminopurine and 5-bromouracil, produce mutations, presumably by being incorporated in replicating DNA in place of their natural counterparts and thereby leading to errors in complement selection. Since these analogues presumably replace their natural counterparts, purine for purine and pyrimidine for pyrimidine, it might be expected that they would be effective in reversing the mutations they induce. In a special class of bacterial virus mutants, investigated by Ernst Freese, this seems to be the case, whereas other classes of mutants, for example those induced by proflavine, are not so reversed by base analogues. It is presumed that the latter brings about replacement of purine by pyrimidine and vice versa. Investigations of this kind offer the hope of giving us a deeper understanding of the mutation process than we now have.

FINE STRUCTURE OF GENES

If, as I have suggested, genes as functional genetic units of DNA are hundreds or thousands of base pairs long, it ought to be possible experimentally to demonstrate multiple mutational sites within a single such unit. Furthermore these should be arranged linearly within a unit. So called fine-structure studies on genes of bacterial viruses, bacteria, fungi and other organisms show that this is indeed the case (Beadle, 1960).

Clearly the unit of mutation is much smaller than the unit of function, for such sites within a functional unit undergo recombination in much the same way as do separate genes within a chromosome.

On the basis of frequencies of such recombination it is possible to construct intragenic maps of mutational sites and thus to show that they are linearly arranged. The shortest distances that can be measured by such recombination are not far from those that are calculated to exist between adjacent nucleotides in DNA.

THE CODING PROBLEM

How are sequences of nucleotides in DNA related to amino-acid sequences in proteins? DNA and protein can be thought of as four- and twenty-symbol systems. If they are both equivalent to simple linear codes, it is clear that it requires at least three nucleotides to specify an aminoacid, for the maximum number of two-letter words is only sixteen. Several possible coding systems have been investigated but so far the problem has not been solved.

Perhaps it will not be until nucleotide sequences in a gene can be compared with amino-acid sequences in the protein specified by that gene. As you know, complete amino-acid sequences have been worked out for only a very few proteins and unfortunately these are from organisms that are far from ideal for genetic study.

There are now investigations under way in many laboratories designed to correlate genetic fine structure with protein fine structure. Virus proteins and bacterial enzymes such as alkaline phosphatase and tryptophane synthetase are among the systems that look especially favourable for such combined genetic and chemical studies.

EVOLUTION

Presumably DNA replication bears no immediate relation to its informational content. Biologically useless DNA is replicated just as faithfully as that playing a vital role. Presumably, too, the mutation process is a random affair. If so, meaningful DNA sequences are much more likely to be made less useful through random mutation than they are to be converted into sequences more useful to the organism of which they are a part. In this sense they may be likened to random typographical errors in a useful message. Direct experiment verifies this expectation; most mutations are unfavourable in the context in which they occur.

If mutations are mostly unfavourable but are as faithfully reproduced as are their normal counterparts, why do they not accumulate with successive rounds of replication just as a typist would accumulate errors in the Lord's Prayer if she were to type copies in succession, each from the previous typing, in a purely mechanical way without proofreading or correcting errors! The answer is that biologically unfavourable mutations specify organisms with reduced reproductive fitness. The errors in their DNA are not corrected but the lines of descent carrying such errors are statistically bred out of existence over a shorter or longer number of generations depending on the degree of reduction in reproductive fitness. We call this natural selection. Positive evolutionary progress depends on rare favourable mutations that increase reproductive fitness. They gradually replace the ancestral types from which they arose. Modern concepts of organic evolution hold that all living systems have evolved gradually, mutation by mutation, from pre-existing organisms. Sometimes, as in parasites, it is advantageous to become simpler and to count increasingly on the host for raw materials, a proper environment and protection against enemies. But from the beginning, the overall trend in many lines such as our own has evidently been toward greater complexity.

Speculation as to how organic evolution began in the first place leads to the conclusion that there is no clear qualitative break in the sequence of events that spans the advent of life on earth. This is only another way of saying that we cannot clearly distinguish between living and non-living systems. No matter what system one contemplates, it is possible to imagine a closely related one, only slightly simpler, that could have given rise to the former by a single mutation-like step.

The simplest living systems we know today are viruses. Tobacco mosaic virus particles are submicroscopic rods about 800 Å units in length. They consist of a cylindrical protein coat and an RNA core, the latter being made up of about 6000 nucleotides. In a living tobacco cell infected with a particle of this virus more viruses are synthesised.

The bacterial virus øX174 contains about the same amount of nucleic acid. Its units are almost spherical. They, too, have a protein coat and a nucleic acid core, but their nucleic acid is DNA, not RNA. Some people say such viruses are not living—that they cannot carry on metabolism, synthesise their component parts, or do any of several other things that living organisms do. But they do reproduce their kind, given a proper environment. And they are mutable, and hence capable of organic evolution. Whether one calls them living or not depends on one's definition of life and that in the end must be purely arbitrary.

Let us imagine that virus-like systems were among the earliest "organisms" to evolve on earth. They were not like present day viruses, for there were no living cells from which they could derive their parts-nucleotides, amino-acids, perhaps a few enzymes, etc. We can reasonably believe that at that stage of the earth's history there were myriads of spontaneously formed organic molecules around, including nucleotides, amino-acids, proteins, etc., and that these were the building blocks of the postulated virus-like creatures. After all we now have abundant experimental evidence that a variety of organic molecules are formed spontaneously under conditions assumed to have been characteristic of parts of the earth's crust a few thousand billion years ago.

If the building blocks of the postulated virus-like systems were around, erations depending on the degree of reduction in reproductive fitness. We call this natural selection. Positive evolutionary progress depends on rare favourable mutations that increase reproductive fitness. They gradually replace the ancestral types from which they arose.

Modern concepts of organic evolution hold that all living systems have evolved gradually, mutation by mutation, from pre-existing organisms. Sometimes, as in parasites, it is advantageous to become simpler and to count increasingly on the host for raw materials, a proper environment and protection against enemies. But from the beginning, the overall trend in many lines such as our own has evidently been toward greater complexity.

Speculation as to how organic evolution began in the first place leads to the conclusion that there is no clear qualitative break in the sequence of events that spans the advent of life on earth. This is only another way of saying that we cannot clearly distinguish between living and non-living systems. No matter what system one contemplates, it is possible to imagine a closely related one, only slightly simpler, that could have given rise to the former by a single mutation-like step.

The simplest living systems we know today are viruses. Tobacco mosaic virus particles are submicroscopic rods about 800 Å units in length. They consist of a cylindrical protein coat and an RNA core, the latter being made up of about 6000 nucleotides. In a living tobacco cell infected with a particle of this virus more viruses are synthesised. The bacterial virus øX174 contains about the same amount of nucleic acid. Its units are almost spherical. They, too, have a protein coat and a nucleic acid core, but their nucleic acid is DNA, not RNA.

Some people say such viruses are not living—that they cannot carry on metabolism, synthesise their component parts, or do any of several other things that living organisms do. But they do reproduce their kind, given a proper environment. And they are mutable, and hence capable of organic evolution. Whether one calls them living or not depends on one's definition of life and that in the end must be purely arbitrary.

Let us imagine that virus-like systems were among the earliest "organisms" to evolve on earth. They were not like present day viruses, for there were no living cells from which they could derive their parts-nucleotides, amino-acids, perhaps a few enzymes, etc.

We can reasonably believe that at that stage of the earth's history there were myriads of spontaneously formed organic molecules around, including nucleotides, amino-acids, proteins, etc., and that these were the building blocks of the postulated virus-like creatures. After all we now have abundant experimental evidence that a variety of organic molecules are formed spontaneously under conditions assumed to have been characteristic of parts of the earth's crust a few thousand billion years ago.

If the building blocks of the postulated virus-like systems were around, would they not have interacted, again spontaneously, to form nucleic acids, proteins, etc.? Nucleotides interact in the

Kornberg system to form DNA capable of replication. Since the role of man in this experiment is merely that of making the conditions more favourable for this particular reaction—that is, increasing its probability—would it not have occurred in his absence when the conditions became right?As every organic chemist knows, organic molecules are formed through the interaction of inorganic molecules—again when the conditions are right.

And as every inorganic chemist knows elements interact to produce inorganic molecules. Nuclear physicists tell us that elements themselves evolve from simpler elements as a result of processes that are both natural and inevitable, given the appropriate circumstances.Thus it is clear that the sequence: hydrogen—helium—beryllium-8—carbon—oxygen—other elements—water—other inorganic molecules—simple organic molecules—more complex organic molecules like nucleotides, aminoacids, and small proteins—nucleic acids capable of replication—nucleic acids protected by protein coats—virus-like systems with protein coats serving catalytic functions—multigenic but subcellular organisms—simple cellular systems like bacteria—autonomous cellular forms like algae—protozoa-multicellular plants and animals and, in our line of descent, man himself—is a natural one that could have arisen by steps no one of which need have been larger than the individual mutational steps we know in today's living systems.

In the beginning there was a universe of hydrogen. How and when it was created—or whether it is and always has been in a steady state of continuous creation—science knows very little. In whatever way the universe began, there must have been built into it from the very beginning the potentiality of essentially unlimited orderly evolution.

DNA AND THE CHEMISTRY OF INHERITANCE

Living organisms are unique, among the known forms of matter, because they are capable of creating their own specific, highly organized structure out of substances taken from far more disorganized surroundings, and can transmit this capability to their offspring.

Perhaps the oldest and most profound theoretical problem in biology is the effort to explain the curious paradox that, despite its unique capability for self-duplication and inheritance, a living organism is nevertheless a mixture of substances which are separately no more possessed of these properties than are the more prosaic molecules that never occur in living cells.

This question has been at the root of a long train of experiments, debates, and speculations that begins in classical times and continues unbroken through the development of present-day "molecular biology."

The basic issues are simply stated. If the component parts of a cell are not themselves alive, whence come the life-properties exhibited by the whole? Apart from the untenable notion of a mystic non-material "vital force" which supposedly animates the otherwise dead substance of the cell, the debate has elicited two main positions:

- There is, in fact, some special cellular component which possesses the fundamental attribute of self-duplication, and which is therefore a "living molecule" and the basic source of the life-properties of the cell.
- The unique properties of life are inherently connected

> with the very considerable complexity of living substance and arise from interactions among its separable constituents which are not exhibited unless these components occur together in the complex whole. In this view, only the entire living cell is capable of selfduplication.

There is at this time a widespread impression that this issue has now been resolved and that the cell does indeed contain a component—DNA— which, according to the theory of the "DNA code," possesses the basic attribute of life, self-duplication, and which guides the activities of all the inheritable processes of the cell.

The importance of this conclusion is self-evident, for it would answer, at last, the basic question of the origin of the unique properties of life, and, if correct, should lead to unprecedented technological control over these properties.

It is appropriate, therefore, to ask what criteria are required to establish that a molecule, such as DNA, is capable of self-duplication, to examine the degree to which the available evidence meets such criteria, and to determine whether the undoubted importance of DNA in the biology of inheritance may be due to some properties other than those attributed to it by current theory.

SOME IMPLICATIONS

The chief conclusion to be derived from the foregoing considerations is that the unique capability of living organisms for self-duplication and inheritance arises from complex multi-molecular interactions among at least several classes of cellular components.

Neither DNA nor any other cellular component is a "self-duplicating molecule" or the "master chemical of the cell"—terms which sometimes appear in current generalizations. There is no evidence from recent investigations of the bio-chemical aspects of genetics which requires abandonment of the conclusion, long established by biological data, that the least complex agent capable of selfduplication is the intact living cell.

The point of view developed here also suggests alternatives to a number of current views of phenomena related to inheritance:

- Various hypothetical schemes for regulating the "activity" of template genes are invoked in order to explain how the DNA code might operate in cellular differentiation and development. Alternatively, it may be suggested that differentiation is associated with the nucleotide sequestration system, tissue-specific changes in DNA snythesis (e.g., at chromosome "puffs") regulating nucleotide levels and thereby governing the size and metabolic character of the cell.
- The relationship between DNA and a species' characteristic sensitivity to ionizing radiation is often interpreted in terms of radiation-induced mutations in genetic templates. Alternatively, radiation-induced inhibition of DNA synthesis, which results in the accumulation of toxic concentrations of free nucleotides, may be regarded as the mediating process. This view is compatible with Sparrow's observation that in certain non-polyploid plants, radiation sensitivity is proportional to cellular DNA content, and is particularly affected by the relative proportions of heterochromatin.
- Sahasrabudhe has suggested that tumor cells may be characterized by marked changes in free nucleotide level resulting from nucleotide sequestration due to DNA synthesis. In this connection, it is also of interest that tumor and other rapidly growing cells exhibit distinctive pathways of oxidative metabolism, which are in turn regulated by free nucleotide levels.
- The theory of the DNA code suggests a seemingly simple explanation for the emergence of the first forms of life from the prebiotic organic environment: that life began with the fortuitous appearance of a "self-duplicating nucleic acid molecule" which then organized the complex chemistry of life around itself. In contrast, the viewpoint developed here suggests

that the primitive function of nucleic acid was to sequester free nucleotides, and thereby regulate overall metabolic activity. Thus, the nucleic acid template is to be regarded as a late development which improved the precision of the earlier modes of regulation of cellular activity, and which is, in any case, incapable of self-duplication.

The theory of the DNA code, if correct, also leads to important expectations regarding the feasibility of technological control over inheritance. If the nucleotide sequence of DNA were indeed a self-sufficient source of the inherited specificity of living things, then it might be possible, in the not too distant future, to synthesize artificial DNA molecules, which, on being introduced into living organisms, would artificially establish new inheritable characteristics quite outside the range of those normally observed.

However, if, as we have concluded, biological specificity is only partly due to DNA, then it is likely that the specificity represented by proteins may impose severe limits on the acceptability, to the cell, of abnormal DNA nucleotide sequences. This view suggests that technological control over biological inheritance may be far less plausible than indicated by inferences drawn from the code theory.

Finally, the viewpoint developed here bears on some fundamental questions regarding the relationship between physical theory and biology. The theory of the DNA code is often regarded as an example of the success with which "modern physics" can solve basic problems about living systems which have eluded the supposedly less critical analyses of classical biologists.

This view leads to the generalization that, despite its obvious complexity, the living cell must be governed by the "laws of physics"—as revealed by analysis of non-living systems—and that the most effective strategy for elucidating its unique properties is to study isolated parts which are sufficiently simple to permit their analyses in physico-chemical terms.

However, several developments in theoretical physics

suggest that the fundamental properties of matter are in better harmony with the view that the unique properties of the cell are derived from interactions of its molecular parts and are inherently incapable of being elucidated by a simple summation of the observed properties of the isolated parts.

This viewpoint is a direct outgrowth of Bohr's theory of complementarity. It may also be closely related to the more recent theory of the "Bootstrap Universe." This theory suggests that the unique properties of the atomic nucleus are inherently associated with its complexity, and are not accountable by the observed properties of fragments isolated from disrupted nuclei.

It may be suggested then, that, if biology is to be guided by the insights into the properties of matter that are afforded by modern physical theory, the role of DNA in the living cell must be viewed as subsumed under the complex properties of the system—the living cell—of which functional DNA is a part. The theory of the DNA code is sometimes epitomized by the statement "DNA is the secret of life," an aphorism which appears increasingly to guide the course of current biological investigations. The viewpoint developed here suggests that biology might be more wisely guided by the aphorism, "Life is the secret of DNA."

Chapter 12

Aspirin Effect on Human Body

In many industrialized nations, the proliferation of health technologies has coincided with increased numbers of epidemiological studies on the risk of injury resulting from medical products and services. Every few months, a previously unknown hazard associated with a commonly used product is reported in the scientific literature or at scientific meetings.

Such reports are often followed by waves of media attention on the dangers in question, accompanied by public and scientific debate about the "Correct" interpretation of these risks. As Feinstein and others pointed out, all too often faulty studies purporting to show causeand-effect relations between common products and severe adverse outcomes gain widespread media attention, further scaring an already suspicious American public.

Media reports tend to concentrate on rare but dramatic hazards and often fail to report more common but serious risks, such as motor vehicle accidents. Some have suggested that this cycle has created a public "epidemic of apprehension"—even about health technology.

On the positive side, the mass media may be a major channel in alerting the public to important and well-documented dangers, such as the carcinogenic potential of asbestos in workplaces, schools and homes, particularly if the scientific and medical literature diffuses more slowly to appropriate decision makers.

In these situations, governments and professional health associations are faced with a dilemma: At what point does sufficient evidence exist to justify public warnings about the potential dangers of specific products that also impart significant health benefits?

Beyond this issue lies another fundamental question: Under what circumstances can the lay and medical press influence professional and consumer behaviour? Technology-related hazards often appear suddenly and under conditions of uncertainty and such situations do not lend themselves practically or ethically to the methodological demands of experiments.

A fair number of studies exist on the modest effects of mass media campaigns targeting well-established risk behaviors, such as smoking and seat belt use. However, few explorations of media-related changes in health product use appear in the literature. One study used time series analysis to observe increases in rates of discontinuation of intrauterine devices (IUDs) and birth control pills following increased news coverage about their adverse effects. Similarly, the Kellogg Company's national television advertising campaign on the role of fibre in preventing cancer was associated with an increase in the market share of Kellogg's and other bran cereals.

Changes in personal habits are of a different character than changes in purchase decisions following warnings about newly discovered hazards of commonly used products. The desired behaviour change represents not the establishment or cessation of a habit, but simply substitution of one product for another closely related product. Marginal changes in behaviour are likely to be somewhat easier to achieve than the elimination of a strongly established routine, or adoption of a completely new one.

This chapter addresses the question of how the media can change consumer behaviour vis-à-vis popular health-related products by focusing on the diffusion of risk information about the relation between use of aspirin and Reye's syndrome, a rare but potentially fatal or severely debilitating illness usually occurring in young children with flu, varicella, or other viruses. First, the epidemiological evidence supporting the aspirin-Reye's syndrome association is reviewed. Second, the actions of several public and private organizations and the key events that played important roles in promoting or impeding public awareness are described.

Third, there is discussion of the relations between the quantity and timing of medical and lay media attention to this issue, other public education activities and decreases in the use of aspirin for children and reductions in disease incidence. Finally, based on previous research in health education, risk perception and changing professional behaviour, some specific characteristics of risk communication (exemplified by the Reye's syndrome case), which may facilitate changes in health behaviors in response to mass communications, are proposed.

EPIDEMIOLOGICAL STUDIES

Reye's syndrome (RS) is a disease of acute encephalopathy, characterized as a viral illness progressing to delirium, fever, convulsions, vomiting, disturbed respiratory patterns, stupor, seizures, or coma. The syndrome occurs most often in children between age 5 and 15 and its effects are independent of race, ethnicity and gender. Epidemics are often correlated with influenza and varicella outbreaks, as well as other viral conditions. The syndrome was first described by R. D. K. Reye in the early 1960s and possible associations between ingestion of salicylates and subsequent development of RS was first suggested as early as 1962.

National surveillance for RS was begun by the Centers for Disease Control (CDC) during the 1973–1974 nationwide outbreak of influenza B. In late 1976, CDC and state health departments intensified surveillance of the syndrome and have continuously maintained surveillance since that time. During the first several years of monitoring, from 250 to 550 cases of RS were reported each year to the CDC, with the largest number of cases occurring during years of influenza A activity. Fatality rates reached 40%, but have declined to 20%–30% more recently.

Between 1980 and 1982, three case-control studies were published that represented the first major investigations in the United States on the link between ingestion of aspirin and subsequent development of Reye's syndrome. A Michigan study indicated that children with RS were more likely to have received aspirin during a viral illness preceding the onset of RS than controls. The authors were among the first to conclude that "aspirin taken during viral illness may contribute to the development of RS. " Studies in Ohio and Arizona similarly observed an association between aspirin use and Reye's syndrome.

Although no extensive warning campaigns were begun prior to the case control findings, the CDC did publish a recommendation in Morbidity and Mortality Weekly Report (MMWR) that "parents should be advised to use caution when administering salicylates to treat children with viral illnesses, particularly chicken pox and influenza-like illnesses". In 1982, the chairman of the CDC Committee on Infectious Disease, Vincent Fulginiti, concluded that the three studies in Michigan, Ohio and Arizona showed a strong association between aspirin use and RS, stating that "it is the consensus of the committee that there is high probability that the administration of aspirin contributes to the causation of RS" and recommending that aspirin not be prescribed for children with chicken pox or influenza.

Publication of the study findings was accompanied by substantial controversy over their scientific validity, resulting in the formation of a U. S. Public Health Service (PHS) task force to investigate the possible link. The PHS conducted a pilot study in 1984 supporting the findings of the earlier studies and a main study from January 1985 to May 1986, which involved 50 pediatric care centers in the United States. The main study included 27 RS cases and 140 controls and found a strong association between salicylate ingestion during an antecedent illness and the onset of RS.

The authors concluded that the risk of RS is related to both exposure and quantity of salicylates ingested and suggested that over 90% of RS cases are associated with salicylates. More recent evidence indicates that there may be a dose-response relation, further validating the previous attributions of a causal relation.

The publication of these epidemiological studies initiated

many years of active controversy about the hypothesized association, as well as further research on the clinical, pharmacological and epidemiological bases of the relation. Many government and private organizations contributed to the public debate about risks, sometimes in opposite directions.

Although a complete picture of the 10-year history of these controversies is impractical, it is useful to examine the major events and organizations involved in public, professional and media communication campaigns. To accomplish this goal, a comprehensive review was conducted of medical and lay media reports identified from computerized searches, published documents from the CDC, weekly pharmaceutical reports and files maintained by one of the most active private organizations involved in the controversy, Public Citizen's Health Research Group (HRG).

MEDICAL AND PHARMACY REPORTS

Following publication of the first epidemiological findings, articles began to appear in medical journals reporting on the studies, analyzing their methods and concluding with tentative advice about prescribing aspirin to children with fever and chicken pox or influenza. The medical news section of JAMA, for example, published an article in March 1982 recommending acetaminophen as a safe and effective alternative to salicylates for these children. Because of the proven benefits of aspirin for arthritis and other conditions in children, there were also vigorous debates on the issue of whether labels should be placed on salicylatecontaining drugs.

For example, in a November 1982 issue of the American Journal of Diseases of Children, a proponent for warnings argued that "even the fragmentary evidence available does provide some biologic plausibility for the association of aspirin to the development of RS". Other physicians disagreed strongly with the concept of an aspirin warning label, calling it "premature and alarmist, " and commenting that the informed pediatric community had not reached a consensus about the possible hazards of aspirin use.

Letters to the editors argued that the studies were flawed

and parents may stop giving aspirin completely if the child has an uncertain diagnosis. Two years later, articles about RS began to appear in pharmacy journals. For example, in March 1984, American Pharmacy published an article describing the PHS pilot study and advising that "the public health would be served if an error is to be made by making it on the side of caution".

After the pilot study was completed in 1986, the same journal concluded there was a high probability that aspirin contributes to the development of RS and urged pharmacists to educate parents about the association and to suggest alternatives for treatment of fever. In late 1984, the American Pharmacy Association, a professional organization registering the community of professional pharmacists, launched a voluntary educational campaign about RS. The campaign included distribution of warning posters to 50,000 members through a weekly newsletter.

GOVERNMENT ACTIVITIES

Publication of the first three studies linking aspirin and RS resulted in public sector activities to assess the evidence and plan future positions. The Food and Drug Administration (FDA) and the CDC, both within the U. S. Public Health Service, were the primary actors in the federal government. The CDC and its staff carried out the most rigorous epidemiological studies and also communicated and interpreted study findings for the media and public. The FDA was responsible for insuring that any important information on risks of aspirin would be communicated both to consumers and health professionals through public statements, information campaigns and aspirin warning labels.

However, the agency faced two competing pressures: one from the Reagan administration and its Office of Management and Budget, which was attempting to reduce regulatory oversight of private industry; and the other, from Congress and citizen groups, which were demanding government-required warnings about suspected risks as expeditiously as possible. The aspirin-RS link was not unknown to the FDA at the time when the three state studies were published. As early as 1976, the FDA reported indications of a possible association. It was not until 1982, however, that the agency took an active role in the controversy.

In early 1982, an FDA working group conducted a direct audit of the raw data of several state health department studies and co-sponsored a workshop to discuss the issue with the CDC and the National Institute of Allergy and Infectious Diseases. At the completion of the meeting, a majority of the scientists concluded that the evidence suggesting an association was sufficiently strong to warrant warning health professionals and parents. The pressure on the FDA began to mount in many quarters.

By late March 1982, the U. S. House of Representatives Subcommittee on Oversight and Investigations had conducted a preliminary investigation of FDA actions. John Dingell, chairman of the subcommittee, was particularly critical of the FDA's failure to comply with CDC recommendations. Richard Schweiker, Secretary of the Department of Health and Human Services (DHHS), announced in June 1982, a directive to the FDA to undertake an educational campaign aimed at health professionals and parents and a requirement that aspirin labeling be changed to advise against its use in children with influenza or chicken pox.

The August FDA Drug Bulletin included this information and served as a further warning to health professionals. The FDA plan also included broadcasting a series of radio public service announcements (PSAs) and sending brochures to 150,000 pharmacists and 100,000 physicians.

However, Secretary Schweiker stopped the proposed rule making concerning aspirin labeling in fall 1982. Several factors probably accounted for this decision. First, on November 9, the American Academy of Pediatrics advised DHHS that labeling of aspirin should be delayed until there was more conclusive evidence of the association.

Second, based on notes obtained during congressional hearings, the chairman of the House Committee on Natural Resources, Agriculture Research and the Environment also implied that DHHS was acting too quickly. So, on November 18, DHHS announced that new studies were necessary to solve the dispute and formed a task force to examine the evidence further.

The public information campaign, however, did get under way in late 1982, when the Surgeon General issued a newspaper column on the association to 8,000 news outlets. In addition, 673,000 copies of an FDA question and answer brochure were distributed to pharmacies and primary care physicians. The Assistant Secretary for Health also sent letters to physicians enclosing a brochure on RS.

Other potentially more powerful components of the education campaign did not have a smooth beginning. In fall 1983, for example, about half a million FDA-produced pamphlets were to have been distributed to 4,200 supermarkets warning parents of the association, but distribution was banned by the Secretary of DHHS and the pamphlets remained in a warehouse, due in large part to strong lobbying and threatened lawsuits by an aspirin industry-financed organization of pediatricians, the Committee on the Care of Children (CCC), which was created to counteract the warning campaign.

An October 1983 letter from the CCC's attorney outlined the industry's charge that the supermarket pamphlet was misleading in implicating aspirin and salicylates as causes of RS and specifically called for a halt to the publicity campaign and recalls of distributed materials

. Distribution of a 30-second radio announcement to 5,000 radio stations was also canceled. Television PSAs, however, had already been sent to approximately 800 commercial television stations in fall 1983. During this time, a newspaper column prepared by the FDA was distributed to approximately 1,500 newspapers, as well as a new 60second PSA to 200 television stations. The information in the approved materials was not fundamentally different from previous FDA messages and thus was not affected by the CCC's actions.

The publication of the 1984 PHS pilot study prompted additional waves of FDA and DHHS activity. The Institute of

Medicine had already reviewed the data and concluded that they revealed a strong association between RS and the use of aspirin. New DHHS Secretary Heckler stated that the study was "not completely conclusive—but its findings do show an association between the use of aspirin and the onset of RS in children and teenagers" with flu and chicken pox. Subsequently, Secretary Heckler arranged for letters on the pilot study to be sent to pediatricians, family physicians, newspapers and broadcast outlets.

In addition, in early 1985, she called for voluntary labeling of aspirin products by the Aspirin Foundation of America. She requested that manufacturers remove any labels recommending aspirin for flu or chicken pox in children and that all aspirin labels contain information on the possible association between aspirin and RS and a recommendation that aspirin not be used in these circumstances without consulting a physician.

The extent of voluntary compliance by the aspirin industry was controversial. In November 1985, due in part to the pressure from public interest organizations and several U. S. senators and a realization that the voluntary programme was not sufficient, a regulation was approved by Secretary Heckler requiring all salicylate-containing over-the-counter (OTC) medications to be labeled beginning in June 1986. The warning label required the following statement: "Warning;: Children and teenagers should not use this medication for chicken pox or flu symptoms before a doctor is consulted about Reye's syndrome, a rare but serious illness".

This warning was required to precede any additional warning that may appear. The FDA public information campaign continued, with a focus on parents of older children and teenagers who were likely to be self-medicating. The campaign included posters for physician's offices, store posters, PSAs for radio and TV and new brochures and letters for the public and health professionals.

In June 1988, the FDA requirement for RS warning was made permanent and a rule was passed requiring that manufacturers place "attentiongetting statements on the principal display panel" of packaging to notify consumers about the new warning.

ASPIRIN INDUSTRY CAMPAIGNS

Pharmaceutical companies with large shares of the aspirin market played a major role in helping to shape the public debate about the aspirin-RS controversy. Aspirin industry representatives met many times with the FDA and submitted several analyses to federal officials involved in the labeling and dissemination process. Their earliest efforts were to critically evaluate the case control studies and to publish reports questioning their conclusiveness.

In March 1982, Plough, Inc. sent letters to pediatricians questioning the validity of recent reports and news stories linking aspirin to RS and recommended continuing to prescribe aspirin for the reduction of fever in children. Other industry-sponsored reports were published as articles or letters in medical journals. For example, the June 1982 issue of Pediatrics contained an industry-sponsored report suggesting that prodromal illness severity was a confounding cause of increased aspirin use in children with RS.

In November 1982, the CCC filed suit in Federal District Court in Boston seeking a temporary restraining order to prevent the FDA from implementing its public information campaign. Later that month, the District Court in Boston denied the petition for a restraining order and in March 1983, the CCC withdrew its lawsuit. Later that year, the CCC submitted a citizens' petition to the FDA requesting it to issue regulations governing publicity and asserting that the public education campaign on RS and aspirin was inappropriate. A representative of the CCC met with Assistant Secretary Edward Brandt in an attempt to stave off the publicity campaign. In a letter soon after the meeting, he stated:

We believe that the massive Government publicity campaign is based on nothing more than sheer speculation and the inability to admit previous error. We believe that the publicity has caused and will continue to cause needless panic, loss of confidence of patients in their physicians, delayed diagnosis of RS and additional fatal overdose problems related to other medications.

We ask that the present publicity campaign be halted and that materials which have already been distributed be recalled because of the hazard to the public. Instead, we believe that a public service announcement which calls attention to the need for early diagnosis of Reye Syndrome together with the caution relative to the use of all antipyretic medication would be most appropriate.

In October 1983, the CCC sent letters to all commercial television stations, claiming that equal time and the fairness doctrines would apply if PHS RS ads were run. The organization also sent letters to supermarkets who were to receive the new question and answer brochures, advising them that displaying the brochure would be a violation of FDA labeling requirements.

The organization also sent a press release entitled "Doc;tors Counter FDA Campaign—Equal; Time Request Threatened, " which claimed that most scientific experts agreed that early studies were seriously flawed and according to their legal counsel, the FDA was providing misleading and deceptive information. Other industry representatives claimed that the public education programme would contain information on the aspirin-RS association and would thus bias the case-control study designed by the PHS.

Industry representatives also attempted to obtain pre-publication raw data from the CDC pilot study for the stated purposes of determining the validity of study conclusions and defending itself against lawsuits brought by parents of alleged RS victims.

After the pilot data (corroborating the early findings of an aspirin-RS link) were submitted to the prestigious New England Journal of Medicine for possible publication, the Plough, Inc. director of clinical affairs wrote NEJM's editor, Dr. Arnold Relman, calling for a more complete and objective "review of all scientific data underlying this Pilot Study which we believe may be seriously flawed". The letter described a number of potential biases in the study and requested they be considered "so that a balanced presentation of this subject matter is available through the New England Journal of Medicine"). Eventually, in 1986, Plough, Inc. obtained a subpoena allowing its scientists to examine raw data from the governmentsponsored study provided that the identities of study participants be excluded.

In January 1985, after Secretary Heckler's call for voluntary labeling of aspirin-containing products, the policy of the aspirin industry began to change gradually. The first to respond was Schering-Plough, Inc. Members of the Aspirin Foundation followed. The factors responsible for these changes in industry policy are not fully understood. The credibility bestowed on the claims of warning proponents resulting from the publication of the pilot study probably accounted for some of the softening in the industry's position.

A Boston Globe interview with Dr. Joseph White, the president of the Aspirin Foundation, reported his statement that although the government had not yet proven an association between aspirin and RS, the foundation was agreeing to voluntary labeling "to protect those at risk, create some peace and quiet and get some people off the backs of the industry".

The programs, negotiated in meetings with the FDA, included television public service announcements, store campaigns with posters and signs distributed by manufacturers' representatives and label changes. Some large retailers also planned their own public education activities, including posters and signs in the nonprescription drug sections of pharmacies and food stores.

In 1985, more than 800,000 warning posters were distributed and radio and TV PSAs were prepared under the auspices of the Aspirin Foundation. In October, the FDA announced that the voluntary relabeling and educational effort was working well and that 68% of children's aspirin on store shelves by the first week of November had the new labeling.

Drug store marketing services also contained information on the warning poster and a letter from the FDA commissioner urging cooperation with the voluntary programme. Although not all manufacturers were equally active in promoting the educational messages, clearly there had been a softening of the aspirin industry's position following the pilot study. This shift in attitude continued and was reinforced by the publication of the final CDC study in 1987.

CHANGES IN THE INCIDENCE OF REYE'S SYNDROME

By 1987, the incidence of Reye's syndrome had declined to its lowest level since monitoring began in the mid-1970s. The FDA reported in the October 1987 FDA Drug Bulletin: "It appears that the rate of RS has decreased markedly in the U. S., probably as a result of PHS and voluntary industry publicity and educational efforts and that parents are heeding warnings— including those from health professionals and that on the aspirin product labeling—to avoid giving aspirin to children and teenagers with symptoms of chicken pox or flu".

Because of the large number of organizations involved, it is difficult to know with certainty what elements or groups were most effective in changing public and professional behaviour to reduce the incidence of RS. A quantitative analysis of the timing and intensity of communications from government, the lay and medical press, industry and other organizations is crucial to an understanding of the changes that occurred.

Although causeand-effect relations cannot be shown by such analyses, it is informative to isolate the "turning point" in RS incidence and compare it to periods of increased intensity in specific types of public communications.

The treatment of influenza and chicken pox frequently involves advice from a family physician or pharmacist, as well as independent treatment decisions by parents. If the campaign was successful, then it would have raised the awareness of both health professionals and parents concerning the aspirin-RS link. Several studies have documented a congruence between the government and public health messages and physician knowledge and behaviour.

For example, G. L. Rahwan and R. G. Rahwan reported

that 91% of pediatricians and 98% of pharmacists no longer recommended aspirin for children with fever or pain and an almost identical proportion instead recommended acetaminophen. Similarly, an FDA research group used pharmaceutical marketing data to demonstrate that physician prescribing of aspirin to children declined significantly from 1980 to 1985, while acetaminophen prescriptions rose. This effect was not significant for adults, suggesting a selective effect of the publicity and warnings on aspirin use in children.

Other studies suggest that parents became aware of the aspirin-RS association during the 1980s. For example, in 1985, Morris and Klimberg conducted a national telephone survey of 1,155 parents of children under 20 to determine medication use during episodes of influenza or chicken pox.

Fifty-three percent of parents surveyed were aware of the contraindications against aspirin use, 40% could spontaneously recall the name RS and 84% had heard of RS based on a recognition test. Among the group of children who had chicken pox, about 58% of their parents said they gave their child a medication; of these medicines, 54% were nonaspirin products and only 6% were aspirin. Similar figures were also found for children with influenza.

Although the aforementioned studies suggest that consumer and professional knowledge and behaviour had improved by 1985, it is not clear which public and professional communications were the catalysts for these changes. Because many of the aspirin labeling requirements and federally sponsored education programs were delayed, it was hypothesized that the lay and professional media were the primary mechanisms for educating physicians and parents in the early years of this controversy.

In order to better understand these temporal relations, time series measures of Reye's syndrome incidence and the quantity of lay and medical press reports on this topic were constructed, as well as indicators of the timing of major government and private actions or educational programs that might have affected parental and professional behaviour directly or indirectly by stimulating media coverage. Data on the estimated yearly incidence of RS per 100,000 population less than age 18 .from 1977 (when expanded reporting was initiated) to 1989 were obtained from the CDC RS surveillance system. One threat to validity of these data is that increased media attention may itself have resulted in increased case detection over time. Fortunately for this analysis, however, such an effect would result in positive trends in RS incidence. This effect would, if anything, mask a true association between the quantity of media reports and the incidence of RS.

Any article that included "aspirin" or "salicylates" and "Reye;'s Syndrome" in the headlines, text, or descriptors was counted as one "warning" on aspirin and RS. Inspection of a large sample of medical and lay reports confirmed that virtually all mentioned the possible role of aspirin in causing some cases of RS. The previous indicators were not meant to represent all media, because radio and television were also powerful influences on attitudes and beliefs concerning the aspirin-RS link. However, previous studies have indicated that television coverage usually closely parallels newspaper coverage.

Index

N

O

P

R